THE ONE DEVICE

THE ONE DEVICE

THE SECRET HISTORY OF THE iPHONE

BRIAN MERCHANT

Little, Brown and Company

New York • Boston • London

Little, Brown and Company
Hachette Book Group
1290 Avenue of the Americas, New York, NY 10104
littlebrown.com

First Edition: June 2017

Little, Brown and Company is a division of Hachette Book Group, Inc. The Little, Brown name and logo are trademarks of Hachette Book Group, Inc.

The publisher is not responsible for websites (or their content) that are not owned by the publisher.

The Hachette Speakers Bureau provides a wide range of authors for speaking events. To find out more, go to hachettespeakersbureau.com or call (866) 376-6591.

Photograph credits: p. vi courtesy of iFixit, the free repair manual (ifixit.com); p. 35 courtesy of *Harper's Bazaar* (public domain); pp. 39, 42, 353, 356 courtesy of the U.S. Patent and Trademark Office (public domain); p. 41 courtesy of *Punch* (public domain); p. 44 by Vannevar Bush, courtesy of *The Atlantic* (public domain); p. 110 courtesy of MIT Libraries (public domain); p. 125 © David Luraschi, used with permission; p. 140 © TechInsights, Inc., used with permission; p. 172 courtesy of the British Library (public domain); all others provided by the author.

ISBN 978-0-316-54616-4 (hardcover) / 978-0-316-55598-2 (int'l paperback)
LCCN 2017936790

10 9 8 7 6 5 4 3 2 1

LSC-C

Printed in the United States of America

For Corrina, the one person who made this all possible, and Aldus, who will hopefully read it one day, on whatever comes after the iPhone

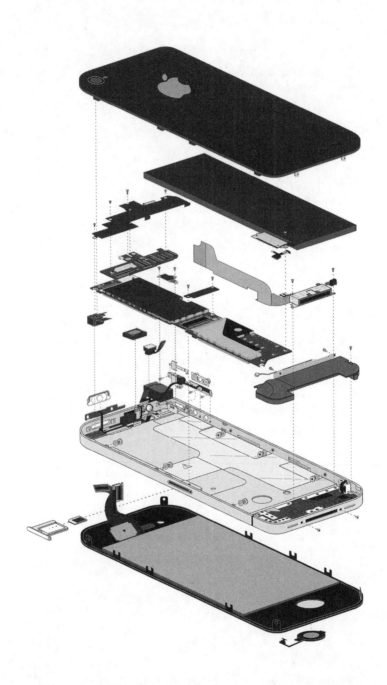

Contents

THE ONE DEVICE

Teardown

On January 9, 2007, Steve Jobs strode onstage at Macworld wearing his trademark black turtleneck, blue jeans, and white sneakers. Twenty minutes into his annual keynote speech, Jobs paused as if to gather his thoughts.

"Every once in a while, a revolutionary product comes along that changes everything," he said. "Well, today, we're introducing three revolutionary products of this class. The first one is a wide-screen iPod with touch controls. The second is a revolutionary mobile phone. And the third is a breakthrough internet communications device. An iPod, a phone, and an internet communicator . . .

"Are you getting it? These are not three separate devices; this is one device, and we are calling it iPhone.

"Today," he added, "Apple is going to reinvent the phone."

Then it did.

★ ★ ★

"Where are. You from?"

"California. Loose angels. Holly would. And. You? Where are. You from? Shang hi?"

Nearly ten years after Jobs's promise, I'm barreling down the highway between the Shanghai Pudong airport and the city's commercial district, and the cabdriver keeps passing his device back to me over the plastic partition. We take turns speaking into a translation app.

"No. Not Shang hi. Hang zoo."

Smog softens Shanghai's neon glow. From here, the skyline is a *Blade Runner* screenshot: brilliant, contorted skyscrapers gracefully recede into a polluted haze.

Our digitally stilted but mostly comprehensible conversation veers from how his night's been (okay) to how long he's been driving a cab (eight years) to the economic conditions of the city (getting worse).

"The pry says keep rising. But. The way jays stay. Flat," a Siri-ish robo-woman's voice intones. When the cabdriver gets fired up, he slows down to crawl, still smack in the middle of the freeway, and cars speed by us as I clutch my seat belt.

"No where. To live." I nod, and he speeds up again.

Interesting, I thought, that my first conversation in Shanghai—where tens of thousands of skilled workers assemble and export the iPhone—was also made possible by the iPhone.

For hours, I'd been dying just to *use* my iPhone—a two-leg transatlantic flight with no wi-fi left me itchy and anxious; my phone was a simmering black hole in my pocket. You might know the feeling: that burning absence whenever you leave it at home or get stuck without a signal. Today, the phone is a lifeline. I felt like I *had* to find a connection to FaceTime with my wife and two-month-old son back home. Not to mention to catch up on email, Twitter, the news, everything else.

How did this happen? How did this one device become the new center of gravity, running our daily lives and enabling things that seemed like science fiction a decade ago—like putting a universal language translator in our pocket? How did it become the one device we need above all?

I was in Shanghai as part of yearlong effort to find out.

* * *

Radical, civilization-scale transformations aren't usually rapid *and* seamless. They tend to be one or the other. But smartphones took over the world quietly and completely in a matter of years, and we barely noticed. We went from having a computer in the household or at work to carrying one everywhere, along with internet access, live chat rooms, interactive maps, a solid camera, Google, streaming video, a near-infinite selection of games, Instagram, Uber, Twitter, and Facebook—platforms that reorganized how we communicate, earn, recreate, love, live, and more—in about the time span of two presidential terms. *We* is the U.S. population, in which smartphone ownership rose from about 10 percent in 2007 to 80 percent in 2016.

That transformation has turned the iPhone into the biggest star of the consumer-electronics world since—scratch that; it's the star of the entire *retail* world. No, even that undersells it. The iPhone might actually be the pinnacle product of all of capitalism to this point.

In 2016, Horace Dediu, a technology-industry analyst and Apple expert, listed some of the bestselling products in various categories. The top car brand, the Toyota Corolla: 43 million units. The bestselling game console, the Sony PlayStation: 382 million. The number-one book series, Harry Potter: 450 million books. The iPhone: 1 *billion*. That's nine zeros. "The iPhone is not only the bestselling mobile phone but also the bestselling music player, the bestselling camera, the bestselling video screen and the bestselling computer of all time," he concluded. "It is, quite simply, the bestselling product of all time."

It's in contention for most stared-at, too. According to Nielsen, Americans spend 11 hours a day in front of their screens. One estimate says 4.7 hours of that time is spent on their phones. (Which leaves approximately 5 waking hours for more traditional life-living: eating, exercising, and driving between the places where we look at screens.) Now 85 percent of Americans say that mobile devices are a central part of their everyday life. You may know you use your phone a lot, but according to a study conducted by British psychologists, you probably use it twice as much as you think you do. Which makes sense, given that we rarely let our phones leave our side—an attachment we've had with very few technologies. "It's almost vanishingly rare that we pick a new device that we always have with us," the historian of mobile technology Jon Agar says. "Clothes—a Paleolithic thing? Glasses? And a phone. The list is tiny. In order to make that list, it has to be desirable almost universally."

Meanwhile, that universally desirable device has made Apple one of the most valuable companies on the planet. What the tech press called the "Jesus phone" now accounts for two-thirds of the company's revenue. Profit margins on the iPhone have been reported to be as high as 70 percent, and as "low" as 41 percent. (No wonder Google's Android phones, which are now even more ubiquitous than iPhones, emulate them so closely they kicked off the industry's most vicious patent battle.) In 2014, Wall Street analysts attempted to identify the world's most profitable product, and the iPhone landed in the top slot—right above Marlboro cigarettes. The iPhone is more profitable than a relentlessly marketed drug that physically addicts its customers.

There's a dependency parallel here too. Like so many, I read news on my iPhone. I'm lost without Google Maps at my fingertips. I keep a constant side-eye on the device for notifications. I check Twitter and Facebook and chat on Messages. I write emails, coordinate workflow, scan pictures. As a journalist, I record interviews and snap publication-quality photos.

The iPhone isn't just a tool; it's the foundational instrument of modern life.

★ ★ ★

So how, and why, did I wind up in Shanghai, looking for the soul of the iPhone? It started months before—when I broke one (again). You know the drill. It slides out of your pocket, and a tiny crack branches into a screen-coating spiderweb.

Instead of buying a new one (again), I decided to take the opportunity to learn how to fix it myself—and find out what's behind the display. I'd been carrying this thing around for years, and I had no idea. So I headed to iFixit HQ, nestled in sleepy San Luis Obispo on the California coast. The company publishes gold-standard gadget-repair guides, and a mellow repair pro named Andrew Goldberg is its lead teardown engineer.

Soon, I was wielding its custom-made tool, the iSclack—picture a pair of pliers whose ends are fixed with suction cups—like a first-year med student. My iPhone 6 and its cracked screen sat between its jaws. I was hesitating: If I popped too hard, I could sever a crucial cable and kill my phone entirely.

"Do it quick," Goldberg said, and pointed to the device, which was beginning to pull apart from the suction. The lights of the workshop studio glared. I was actually sweating. I shuffled my feet, steadied myself, and *snap*—my trusted personal assistant cracked open safely, like the hood of a car.

"That was easy, right?" Goldberg said. It was.

But the relief was fleeting. Goldberg disconnects that cable and removes the aluminum plating inside. Soon my phone's guts are laid out before me on a stark little gadgetry operating table. And if I'm being honest here—it makes me weirdly uncomfortable, sort of like I'm looking at a corpse in the morgue. My iPhone, my intimately personalized life navigator, now looks like standard-issue e-waste—until you know what you're looking at.

The left side of the unit is filled with a long, flat battery—it takes up half the iPhone's real estate. The logic board, a bracket-shaped nest that houses the chips that bring the iPhone to life, wraps around to the right. A family of cables snake around the top.

"There's four different cables connecting the display assembly to the rest of the phone," Goldberg says. "One is going to be the digitizer that receives your touch input. So that works with a whole array of touch capacitors embedded in the glass. You can't actually see them, but when your finger's there…they can detect where you're touching. So that's its own cable, and the LCD has its own cable, the fingerprint sensor has its own cable. And then the last cable up here is for the front piece and camera."

This book is an attempt to trace those cables—not just inside the phone, but around the world and throughout history. To get a better, more tactile sense of the technologies, people, and scientific breakthroughs that culminated in a device that's become so ubiquitous, we take it for granted. The iPhone intertwines a phenomenal number of prior inventions and insights, some that stretch back into antiquity. It may, in fact, be our most potent symbol of just how deeply interconnected the engines that drive modern technological advancement have become.

* * *

However, there's one figure and one figure alone who looms over the invention of the iPhone: Steven P. Jobs, as his name is listed, usually first, on Apple's most important patents for the device. But the truth is that Jobs is only a small part of its saga.

"There was an extraordinary cult of Jobs as seemingly the inventor of this world-transforming gizmo, when he wasn't," the historian David Edgerton says. "There is an irony here—in the age of information and the knowledge society, the oldest of invention myths was propagated." He's referring to the Edison myth, or the myth of the lone inventor—the notion that after countless hours of toiling, one man can conjure up an invention that changes the course of history.

Thomas Edison did not invent the lightbulb, but his research team found the filament that produced the beautiful, long-lasting glow necessary to turn it into a hit product. Likewise, Steve Jobs did not invent the smartphone, though his team did make it universally desirable. Yet the lone-inventor concept lives on. It's a deeply appealing way to consider the act of innovation. It's simple, compelling, and seems morally right—the hardworking person with brilliant ideas who refused to give up earned his fortune with toil and personal sacrifice. It's also a counterproductive and misleading fiction.

It is very rare that there's a single inventor of a new technology—or even a single responsible group. From the cotton gin to the lightbulb to the telephone, most technologies are invented simultaneously or nearly simultaneously by two or more teams working entirely independently. Ideas really are "in the air," as patent expert Mark Lemley puts it. Innumerable inventive minds are, at any given time, examining the cutting-edge technologies and seeking to advance them. Many are working as hard as our mythic Edisons. But the ones we consider iconic inventors are often those whose version of the end product sold the best in the marketplace, who told the most memorable stories, or who won the most important patent battles.

The iPhone is a deeply, almost incomprehensibly, collective achievement. As the pile of phone guts on the iFixit table attests, the iPhone is what's often called a convergence technology. It's a container ship of inventions, many of which are incompletely understood. Multitouch, for instance, granted the iPhone its interactive magic, enabling swiping, pinching, and zooming. And while Jobs publicly claimed the invention as Apple's own, multitouch was developed decades earlier by a trail of pioneers from places as varied as CERN's particle-accelerator labs to the University of Toronto to a start-up bent on empowering the disabled. Institutions like Bell Labs and CERN incubated research and experimentation; governments poured in hundreds of millions of dollars to support them.

But casting aside the lone-inventor myth and recognizing that thousands of innovators contributed to the device isn't enough to see how we arrived at the iPhone. Ideas require raw materials and hard labor to become inventions. Miners on almost every continent chisel out the hard-to-reach elements that go into manufacturing iPhones, which are assembled by hundreds of thousands of hands in city-size factories across China. Each of those laborers and miners are an essential part of the iPhone's story—it wouldn't be in anyone's pocket today without them.

All of these technological, economic, and cultural trends had to glom together before the iPhone could finally allow us to achieve what J.C.R. Licklider termed *man-computer symbiosis*: A coexistence with an omnipresent digital reference tool and entertainment source, an augmenter of our thoughts and enabler of our impulses. The better we understand the complexity behind our most popular consumer product and the work, inspiration, and suffering that makes it possible, the better we'll understand the world that's hooked on it.

None of this diminishes the achievements of the designers and engineers at Apple who ultimately brought the iPhone to market.

Without their engineering insights, key designs, and software innovations, the immaculately crafted one device may never have become just that. But thanks in part to Apple's notorious secrecy, few even know who they are.

That culture of secrecy extends to the physical product itself. Have you ever tried to pry your own iPhone open to look inside? Apple would rather you didn't. One of the ways it became the most profitable corporation on the planet is by keeping us out of the morgue. Jobs told his biographer that allowing people to tinker with his designs "would just allow people to screw things up." Thus, iPhones are sealed with proprietary screws, called Pentalobes, that make it impossible to open up your own phone without a special tool.

"I've had people say, 'Oh, my phone, it doesn't last as long as it used to,'" iFixit CEO Kyle Wiens says. "And I'm like, 'You know you can swap the battery.' I swear to God I've had people say, 'There's a battery inside my phone?'" With the rise of slick, sealed smartphones, we risk happily swiping along as, to borrow Arthur C. Clarke's famous line, this sufficiently advanced technology becomes indistinguishable from magic.

So let's open the iPhone up, to discover its beginnings and evaluate its impact. To bust the Jobs-Edison myth of the lone inventor. To understand how we got to iPhone.

Toward that end, I traveled to Shanghai and Shenzhen, and snuck into the factory where Chinese workers piece the phones together, where a suicide epidemic exposed the harsh conditions in which iPhones get made. I had a metallurgist pulverize an iPhone to find out exactly which elements lay inside. I climbed into mines where child laborers pry tin and gold from the depths of a collapsing mountain. I watched hackers pwn my iPhone at America's largest cybersecurity conference. I met the father of mobile computing and heard his thoughts about what the iPhone meant for his dream. I traced the origins of multitouch through a field of unacknowledged pioneers. I interviewed the transgender chip designer who made the iPhone's brain possible. I met the unheralded software-design geniuses who shaped the iPhone's look and feel into what it is today.

In fact, I spoke to every iPhone designer, engineer, and executive who was willing to do an interview. My goal is that by the end of this book, you'll glance into the black mirror of your iPhone and see, not the face of Jobs, but a group picture of its myriad creators—and have a more nuanced, true, and, I think, compelling portrait of the one device that pulled us all into its future.

A quick note on Apple: Investigating the iPhone is a paradoxical task. A parade of pundits, anonymous sources, and blog posts offer an endless stream of opinions about everything Apple does. A few flat cryptic words from an Apple press release often provides the "official" record. Apple grants next to no interviews with its staff, and the journalists who do get interviews are typically handpicked for their long-standing or friendly relationships with company. I am not one of those journalists—in fact, I'm not even much of a gadget geek. (While I've been on the science and tech beat for about a decade now, I've spent more time covering oil spills than product demos.) So, while I made Apple officials fully aware of this project from the outset and repeatedly spoke with and met their PR representatives, they declined my many requests to interview executives and employees. Tim Cook never answered my (very thoughtful) emails. To tell this story, I met current and former Apple employees in dank dive bars or spoke with them over encrypted communications, and I had to grant anonymity to some of those I interviewed. Many people from the iPhone team still working at Apple told me they would have loved to participate on the record—they wanted the world to know its incredible story—but declined for fear of violating Apple's strict policy of secrecy. I'm confident that the dozens of interviews I did with iPhone innovators, the talks I had with journalists and historians who've studied it, and the documents I've obtained about the device all helped render a full and accurate portrait.

That portrait will emerge on two tracks. The first puts you inside Apple to show how the iPhone was imagined, prototyped, and created by a host of unsung innovators—those who pioneered new ways of manipulating and interacting with information. These four sections start when you turn the page, and appear throughout the middle and at the end of the book—fitting, I think, as countless people gave rise to the one device, but Apple ultimately made the iPhone.

The second will follow my efforts to uncover the iPhone's raw source material, to meet with the minds and hands that made it possible around the globe. These chapters will start at Chapter 1, and proceed from examining the century-old origins of the idea of a "smartphone" to exploring the powerful technologies gathered under its hood, to investigating how all those parts are assembled in China, to visiting the black markets and e-waste pits where they ultimately wind up.

With that, let's make our first stop—Apple HQ in Cupertino, California, the heart of Silicon Valley.

i: Exploring New Rich Interactions

iPhone in embryo

Apple's user-testing lab at 2 Infinite Loop had been abandoned for years. Down the hall from the famed Industrial Design studio, the space was divided by a one-way mirror so hidden observers could see how ordinary people navigate new technologies. But Apple didn't do user testing, not since Steve Jobs returned as CEO in 1997. Under Jobs, Apple would show consumers what they wanted, not solicit their feedback.

But that deserted lab would make an ideal hideaway for a small group of Apple's more restless minds, who had quietly embarked on an experimental new project. For months, the team had held unofficial meetings marked by freewheeling brainstorms. Their mission was vague but simple: "Explore new rich interactions." The ENRI group, let's call it, was tiny. It included a few of Apple's young software designers, a key industrial designer, and a handful of adventurous input engineers. They were, essentially, trying to invent new ways of interfacing with machines.

Since its inception, the personal computer had relied on a century-old framework that allowed humans to tell it what to do: A keyboard laid out like a typewriter, the same basic tool a nineteenth-century newspaperman used to write copy. The only major addition to the input arsenal had been the mouse. Throughout the information revolution of the second half of the twentieth century, that was how most people navigated its bounty — with a typewriter and a clicker. Near-infinite digital possibilities, dusty old user interface.

By the beginning of the twenty-first century, the internet was mainstream and maturing. Online media was complex and interactive. Apple's own iPod was moving digital music into people's pockets, and the personal computer had become a hub for maps, movies, and images. The ENRI group predicted that typing and clicking would soon prove frustratingly cumbersome, and we'd need new ways to interact with all that rich media — especially on Apple's storied computer. "There was a core little secret group," says one member, Joshua Strickon, "with the goal of re-envisioning input on the Mac."

The team was experimenting with every stripe of bleeding-edge hardware — motion sensors, new kinds of mice, a burgeoning technology known as multitouch — in a quest to uncover a more direct way to manipulate information. The meetings were so discreet that not even Jobs knew they were taking place. The gestures, user controls, and design tendencies stitched together here would become the cybernetic vernacular for the new century — because the kernel of this clandestine collaboration would become the iPhone.

Yet its pioneers' achievements have largely been hidden from view, stranded on the other side of that one-way mirror, by an intensely secretive corporation and its late spotlight-commanding CEO. The story of the iPhone starts, in other words, not with Steve Jobs or a grand plan to revolutionize phones, but with a misfit crew of software designers and hardware hackers tinkering with the next evolutionary step in human-computer symbiosis.

Assembling the Team

"User-interface design is still unknown to most people, even now," a member of the original iPhone team tells me. For one thing, the term *user interface* feels pulled right from a technical manual; the label itself seems uniquely designed to dull the senses. "There's no rock-star UI designer," he says. "There's no Jony Ive of UI." But if there were, they'd be Bas Ording and Imran Chaudhri. "They're the Lennon-McCartney of UI."

Ording and Chaudhri met during some of Apple's darkest days. Ording, a Dutch software designer with a flair for catchy, playful animations, was hired into the Human Interface group in 1997, the year the company hemorrhaged a billion dollars in lost revenue and Jobs returned to stanch the bleeding. Chaudhri, a sharp British designer as influenced by MTV's icons as Apple's, had arrived a few years before and survived Jobs's slash-and-burn layoffs. "I first met Imran at some point in the parking lot smoking cigarettes," Ording says. "We were like, 'Hey, dude!'" They make for an odd pair; Bas is lanky, easygoing, and almost preternaturally good-natured, while Imran is intense, fashionable, and broadcasts an austere cool. But they hit it off right away. Soon, Ording convinced Chaudhri to come over to the UI group.

There, they joined Greg Christie, a New Yorker who'd come to Apple in 1995 for one reason: "To do Newton"—Apple's personal digital assistant, an early stab at a mobile computer. "My family thought I was nuts moving to Apple," he says, "working for this company that's clearly going out of business." The Newton didn't sell well, so Jobs killed it, and Christie wound up in charge of the Human Interface group.

As Jobs placed renewed focus on Apple's flagship Macs, Bas and Imran cut their teeth updating the look and feel of its age-worn operating system. They worked on pulsating buttons, animated progress bars, and a glossy, transparent look that rejuvenated the appeal of the Mac. Their creative partnership bloomed. They helped prove that user interface design, long derided as dull—the province of grey user settings and drop-down menus; "knobs and dials" as Christie puts it—was ripe for innovation. As Bas and Imran's stars rose inside Apple, they started casting around for new frontiers.

Fortunately, they were about to find one.

* * *

While training to be a civil engineer in Massachusetts, Brian Huppi idly picked up Steven Levy's *Insanely Great*. The book documents how in the early 1980s Steve Jobs separated key Apple players from the rest of the company, flew a pirate flag above their department, and drove them to build the pioneering Macintosh. Huppi couldn't put it down. "I was like, 'Wow, what would it be like to work at a place like Apple?'" At that, he quit his program and went back to school for mechanical engineering. Then he heard Jobs was back at the helm at Apple—serendipity. Huppi landed a job as an input engineer there in 1998.

He was put to work on the iBook laptop, where he got to know the Industrial Design group, whose profile had already begun to rise under its head, Jonathan Ive. Jobs's streamlining of the company had placed a renewed focus on design, and the ID group's head-turning, color-splashed Bondi Blue iMac—a radical departure for the beige desktop-heavy industry—had helped turn Apple's fortunes around at the end of the '90s. The gig turned out to be a little less insane than Huppi might have imagined, though. His work consisted largely of shipping one laptop, then getting to work updating its successor. He hadn't quit civil engineering to iterate laptop hardware; he was after something more in the pirate flag–flying, industry-shaking vein. So he turned to one industrial designer in particular—Duncan Kerr, who'd worked at the famed design firm IDEO before heading to Apple. "Duncan is the least ID guy of all the ID guys," Chaudhri says—someone who was as interested in what was happening on the screen as what the screen was shaped like.

"We'd been discussing how it'd be really cool to sit down and really focus on talking about a very user-centric approach to what we might do with input," Huppi says. They wanted to reimagine how humans interacted with computing devices from the ground up, and to ask, what, exactly, they wanted those interactions to be. So Kerr approached Jony Ive to see if the ID group could support a small team that met regularly to investigate the topic. Ive was all for it, which was good news—if anyone hoped to embark on a wild, transformative project and actually hope to see it come to fruition, ID was the place to go.

"I knew politically that it had to go through ID," Huppi says, "because they had all the power, and they had Steve's ear."

Huppi knew Greg Christie from a laptop project, and Ording and Chaudhri were already working with Kerr. The chip architecture expert and Newton veteran Mike Culbert, as well as Huppi's boss Steve Hotelling, joined the talks. Then, there was the new recruit— they'd just hired Josh Strickon out of MIT's Media Lab. There, Strickon had spent years enmeshed in experimental fusions of technology and music. For his master's thesis, he concocted a laser range finder for hand-tracking that could sense multiple fingers. "He seemed like one of these guys that had lots of interaction experience," Huppi says, "and I was like, he'd be perfect for this brainstorming group."

★ ★ ★

When Joshua Strickon arrived at Apple in 2003, the company was pocked by uncertainty all over again. The iMac had won accolades and steadied sales, but the tech bubble had burst; profits had fallen, and Apple was losing money for the first time since Jobs's return. The iPod had yet to take off, and the rank and file were anxious. "When I got there," Strickon says, "the stock price was like four- teen dollars and no one had had a raise in who knows how long." Apple placed him in a windowless office with malfunctioning hard- ware. "They had given me a laptop and a desktop," he says, "and the machines were crashing all the time." Meanwhile, the Cuper- tino campus teemed with Apple "fanatics," a number of whom made no secret of their Steve Jobs idolatry. "Apple is kind of a weird place," he says. "You've got people dressing like Steve." There were so many Steve look-alikes, in fact, that Strickon couldn't pick out the real thing. He'd been on the lookout for Jobs since he got to Cupertino—his thesis adviser had worked for Apple years before, and Strickon wanted to say hello on his behalf. But when he found himself next to the CEO in line for a burrito at the cafeteria, he just assumed Jobs was an acolyte. "At the time I didn't realize it was him," Strickon says. "I thought it was just somebody who likes to dress like him."

Strickon was young—he'd finished his PhD and showed up at Apple in his mid-twenties, expecting to find other members of a freshmen class. "But the company was mostly middle-aged men, which was kind of a shocker," he says. He found the atmosphere stifling and buttoned up, which he attributed to all the Jobs worship. "I had friends at Google, and [there it] was like kids running around with no parents." But at Apple, "people didn't feel empowered to have ideas and to follow through.... Everything was micromanaged by Steve," Strickon says.

Strickon's skills—a unique knowledge of touch-based sensors and the software needed to control them, an intrepid musical sensibility, and a flair for experimentation—would translate well to the percolating project, Huppi thought. Even if they wouldn't translate so well to negotiating Apple's corporate culture.

★ ★ ★

So, the key proto-iPhoners were European designers and East Coast engineers. They all arrived at Apple during its messy resurgent years, just before or just after the return of Jobs. They were in their twenties and thirties, ambitious, and keen to experiment with new technologies: Bas Ording, a user-interface wunderkind who took cues from typesetting and gaming; Imran Chaudhri, a hacker-influenced designer who could straddle the gap between SV and MTV; Joshua Strickon, an MIT-trained sensor savant with an ear for electronica and a feel for touchscreens; Brian Huppi, a jack-of-all-trades who could build just about anything; and Duncan Kerr, a decorated designer intent on marrying industrial design to digital interfaces. With support from industry vets like Steve Hotelling and PDA pioneers like Greg Christie, the ENRI meetings would lay out a blueprint for the next generation of mobile computing.

New Rich Interactions

The ENRI project started out with simple brainstorming, in Apple's most hallowed sanctum. With a handful of young men around a conference table, laptops open; with whiteboard drawings and keynote presentations. With extensive note-taking and weekly meetings.

"We almost always met over in the Industrial Design studio, you know, just kicking around ideas all over the map," Huppi says. The central question was simple: "What are the new features that we want to have in our experiences?"

The fact that they were having these conversations at all was a small step forward, as this brand of cross-pollination wasn't common at the time. "What was weird was that you had the Industrial Design group, and mostly what they did was build physical mock-ups," Strickon says. "Nonfunctioning mock-ups, like when you go into a cell phone store and see those plastic models; it'd be like that. They'd spend hours and hours looking at shapes and forms and building weighted versions of that stuff, and it seemed kind of counterproductive because they couldn't see what these things felt like in actual use."

The ENRI meetings aimed to change that, to help infuse that celebrated design work with fully functional input technologies and user interfaces—and to tap into fresh ideas about how they all might work together. "We would just go into that ID studio and, you know, just talk," Huppi says. "This went on for a good six months."

There were a lot of ideas. Some feasible, some boring, some outlandish and borderline sci-fi—some of those, Huppi says, he "probably can't talk about," because fifteen years later, they had yet to be developed, and "Apple still might want to do them someday."

"We were looking at all sorts of stuff," Strickon says, "from camera-tracking and multitouch and new kinds of mice." They studied depth-sensing time-of-flight cameras like the sort that would come to be used in the Xbox Kinect. They explored force-feedback controls that would allow users to interact directly with virtual objects with the touch of their hands.

"Phones," Strickon adds, "weren't even on the table then. They weren't even a topic of discussion." But the ID group fabricated plenty of cell phones. Not smartphones, but flip phones. The husks of stylish phone bodies dotted the design studio. "There were many models of flip phones of various sorts that Apple had been working on," Huppi says. "I mean very Apple-ized, very gorgeous and beautiful, but they were they were basically various takes on cell phones with buttons." (This might explain why Apple had by this point already registered the domain iPhone.org.)

The talks began to gravitate toward a recurring source of the group's frustration. "One of the themes that kept coming up over and over again was the theme of—I call it navigation," Huppi says, "key things like scrolling and zooming."

Key things that people wanted to be able to do with their richer and more interactive media, now that the web was booming and computers were more powerful—things that tested the UI limits of the decades-old mouse-and-keyboard combo. "It really kind of started by listing things like, 'I wish *this* would work better,'" Huppi says. In 2002, if you wanted to zoom in on an image, you had to drag your cursor to a menu, click, select the amount you wanted to enlarge it by, and click again or hit Enter. Scrolling and panning meant still more clicks; finding the tiny scroll-bar ball and dragging it around. Small things, maybe, but performing these actions dozens of times a day was a pain—especially for designers and engineers. Chaudhri, for one, was interested in directly interacting with the screen—streamlining receptive acts like closing windows. "What if you could just tap, tap, tap them and be done?" he says. That sort of direct manipulation could make navigating computers more efficient, expressive, more fun.

Fortunately, there was a consumer technology out there that already allowed users to do something like that, if not quite in the exact form the ENRI crew was after. In fact, one of Apple's engineers was using it. Around that time, Tina Huang had shown up to work with an unusual, plastic black touchpad marketed to computer users with hand injuries. It was made by a small Delaware-based company called FingerWorks. "At the time I was doing a lot of work that involved me testing with a mouse," Huang tells me, including a lot of dragging and dropping. "So I was having some wrist troubles and I think that definitely motivated me to get the FingerWorks."

The trackpad allowed her to use fluid hand gestures to communicate complex commands directly to her Mac. It let her harness what was known as multitouch finger tracking, and, Chaudhri says, it inspired the group to examine the technology.

"We kind of started playing around with multitouch and that was the thing that resonated with a lot of people," Strickon says. He was familiar with the upstart company, and he suggested they reach out.

"I was like, you know, we've actually seen these guys," Huppi says. They'd been in and out of Cupertino taking meetings over the last couple of years, but had never gotten much traction. Finger-Works was founded by a brilliant PhD student, Wayne Westerman, and the professor advising him on his dissertation. Despite generally agreeing that the core technology was impressive, Apple's marketing department couldn't figure out how they would use multitouch, or sell it. "We said, well, it's time to look at it again," Huppi says. "And it was like, Wow, they really have figured out how to do this multitouch stuff with capacitive sensing." It's impossible to understand the modern language of computing, or the iPhone, without understanding what that means.

On Touch

At the time, touch tech was largely limited to resistive screens—think of old ATMs and airport kiosks. In a resistive touchscreen, the display is composed of layers—sheets coated with resistive material and separated by a tiny gap. When you touch the screen with your finger, you press the two layers together; this registers the location of said touch. Resistive touch is often inexact, glitchy, and frustrating to use. Anyone who's ever spent fifteen minutes mashing their fingers onto a flight-terminal touchscreen only to get flickering buttons or random selections is keenly aware of the pitfalls of resistive touch.

Instead of relying on force to register a touch, capacitive sensing puts the body's electrochemistry to work. Because we're all electrical conductors, when we touch a capacitive surface, it creates a distortion of the screen's electrostatic field, which can be measured as a change in capacitance and pinpointed rather precisely. And Finger-Works appeared to have mastered the technology.

The ENRI team got their hands on a FingerWorks device, and found they came with diagrams detailing dozens of different gestures. Huppi says, "I sort of likened it to a very exotic instrument: not too many people can learn how to play it." As it had in the past, Apple's tinkerers saw the chance to simplify. "The core of the idea was there, which was that there were some gestures, like pinch to zoom," Huppi says. "And two-finger scrolling."

A new, hands-on approach to computing, free of rodent intermediaries and ancient keyboards, started to seem like the right path to follow, and the ENRI team warmed to the idea of building a new user interface around the finger-based language of multitouch pioneered by Westerman—even if they had to rewrite or simplify the vocabulary. "It kept coming up—we want to be able to move things on the screen like a piece of paper on the table," Chaudhri says.

It would be ideal for a trackpad as well as a touchscreen tablet; an idea long pursued but never perfected in the consumer market—and one certainly interesting to the vets of the Newton (which had a resistive touch screen) who still hoped to see mobile computing take off.

And it wouldn't be the first time a merry band of Apple inventors plumbed another organization for UI inspiration. In fact, Silicon

Valley's premier Prometheus myth is rife with parallels: In 1979, a young squad of Apple engineers, led by Steve Jobs, visited the Xerox Palo Alto Research Center and laid eyes on its groundbreaking graphical user interface (GUI) boasting windows, icons, and menus. Jobs and his band of "pirates" borrowed some of those ideas for the embryonic Macintosh. When Bill Gates created Windows, Jobs screamed at him for stealing Apple's work. Gates responded coolly: "Well, Steve, I think there's more than one way of looking at it. I think it's more like we both had this rich neighbor named Xerox and I broke into his house to steal the TV set and found out that you had already stolen it."

While the brass set up meetings with the Delaware touch scholars, the ENRI team set to thinking about how they could start experimenting with multitouch in the meantime, and how splicing FingerWorks tech into a Mac-powered device might work. There was, from the outset, a major obstacle to confront: They wanted to interact with a transparent touch*screen* — but FingerWorks had used the technology for an opaque keyboard pad.

The solution? An old-school hardware hack.

Rigged

To find inspiration for a prototype, the team turned to the internet. They found videos of engineers doing direct manipulation, Huppi says, by projecting over the top of an opaque screen. "And we're like, 'This is exactly what we're talking about.'"

They brought in a Mac, set up a projector to hang over a table, and positioned the trackpad beneath it. The idea was to beam down whatever was being shown on the screen of the Mac, so that it'd become a faux 'screen' atop the trackpad. "We basically had this table with a projector over the top, and there's this trackpad, looking like it was like an iPad sitting on the table," Huppi says.

Problem was, it was hard to focus the new "screen." "I literally went home that day and got some crazy close-up lenses out of my garage, and taped them onto the projector," Greg Christie says. The lens did the trick. "If you focus everything the right way, you could actually project an image of the screen onto this thing," Huppi says.

Finally, they needed a display. For that, they went low-tech. They put a white piece of printer paper over the touchpad, and the touchscreen simulation was complete. Clearly, it wasn't perfect. "You got a bit of a shadow from your fingers," Bas Ording says, but it was enough. "We could start exploring what we could do with multitouch."

The Mac/projector/touchpad/paper hybrid worked—barely—but they also needed to customize the software if they were going to experiment with the touch dynamics in earnest, and put their own spin on the interface. That's where Josh Strickon came in. "I was writing a lot of the processing algorithms for doing the finger detection," Strickon says, as well as "glue software" that gave them access to multitouch data generated by the experiments.

Freewheeling as it was, the project nonetheless began to take on a secret air. "I don't remember a day when someone said, 'Okay, we're not allowed to talk about it anymore,'" Strickon says, but that day came. With good reason: The ENRI team's experiment had suddenly become an exciting prospect, but if Steve Jobs found out about it too early and disagreed, the whole enterprise stood to get shut down.

User Testing

The experimental rig was perfect for its new home—that empty old user-testing facility. It was spacious, about the size of a small classroom. Giant surveillance cameras dangled from the ceilings, and behind the one-way mirror, there was a room that looked like an old recording studio, replete with mixing boards. "I'm sure it was all top-of-the-line stuff back in the eighties," Huppi says. "We laughed that the video recording equipment was all VHS!" And they needed security clearance to get in—Christie was one of the few in the company who had access at the time.

"It was kind of a weird space," Strickon says. "The irony was, we were trying to solve problems about the user experience in a user-testing room without ever being able to bring an actual user into it."

Ording and Chaudhri spent hours down there, hammering out demos and designs, building the fundaments of a brand new interface based entirely on touch. They used Strickon's data feed to create tweaked versions of the FingerWorks gestures, and tested new ideas of their own. They homed in on fixing the ENRI group's shit list: Pinch to zoom replaced a magnifying glass icon, a simple flick of the screen simplified click-and-drag scrolling.

Throughout, their creative partnership made for a powerful symbiosis. "Bas was a bit better with the technical side," Chaudhri says, "and I could contribute more with the artistic elements." From an early age, Chaudhri was chiefly interested in how technology was intersecting with culture. "I wanted to be one of three places: the CIA, MTV—or Apple," he says. After interning at Apple's Advanced Technology Group, he was offered a job in Cupertino. His friends were dubious. "You'll spend all your time designing little icons," they said. He laughed them off and took it. "It turns out they'd only be thirty percent right."

His talent with icons, however, made him an ideal fit for Bas's animated experiments. "We worked together quite well," Ording says. "He was doing more icons and nice graphics—he's really good at setting up the whole style. I was a little bit better at doing the interactive prototyping and the feel and dynamic parts." He's being modest, of course. Mike Slade, a former adviser to Steve Jobs, described Ording as a wizard: "He'd take ninety seconds pecking away, he'd hit a button, and there it was—a picture of whatever Steve had asked for. The guy was a god. Steve just laughed about it. 'Basification in progress,' he'd announce." Ording's father ran a graphic design company outside of Amsterdam, and he learned to code as a kid—maybe it was in his blood. Regardless, industry giants like Tony Fadell hail him as a visionary. One of his peers from the iPhone days puts it this way: "I don't know what else to say about Bas, that guy's a genius."

The new touch-based template proved so promising, even exhilarating, that Chaudhri and Ording would pass entire days down there, sometimes without realizing it—UI's Lennon and McCarthy at work.

"We'd go in with the sun and leave with the moon," Chaudhri says. "We'd forget to eat. If you've ever been in love, not having a single care, that's what it was like. We knew it was big."

"There were no windows, it was kind of like a casino," Ording says. "So you could look up and it'd be four o'clock and we'd worked right through lunch."

They started to shield their work from outsiders, even from their boss, Greg Christie—they didn't want anything to intrude on the flow and momentum of the progress. "At that point," Chaudhri says, "we stopped talking to people. For the same reasons that start-ups go into stealth mode." And they didn't want it to get shut down before they could effectively demonstrate the full scope of their gestating UI's potential. Naturally, their boss was irked.

"I remember we were going to Coachella, and Christie told us, 'Maybe when you get back from that orgy in the desert, you can tell me what the hell you're doing down there,'" Chaudhri says.

They cooked up compelling demos that showcased the potential of multitouch: maps you could zoom and rotate, and pictures that you could bounce around the screen with a quick pull of your fingers. They uploaded vacation photos and subjected them to multitouch experiments. "They were the masters of coming up with the UI stuff," Huppi says. People would gather around as Ording used two fingers to rotate and zoom in on globs of color, manipulating the pixels in a smooth, responsive state. Ording and Chaudhri say it was already clear that what they were working on had the potential to be revolutionary.

"Right away, there was something cool about it," Ording says. "You could play with stuff, and drag stuff around the screen, and it would bounce, or you could pinch zoom, all that kind of stuff." You know, the kind of stuff that would become the baseline for a new-fangled mobile machine-human symbiosis.

It was time to put some of that genius to the test.

Showtime

With a handful of working demos and a reasonably reliable rig in place, Duncan Kerr showed the early prototype to Jony Ive and the rest of the ID group. "It was amazing," core member Doug Satzger said, sounding taken aback. Among the most impressed was Ive. "This is going to change everything," he said.

But he held off on sharing the project with Jobs—the model was in a cumbersome, inelegant conceptual state, and he worried that Jobs would dismiss it. "Because Steve is so quick to give an opinion, I didn't show him stuff in front of other people," Ive said. "He might say, 'This is shit,' and snuff the idea. I feel that ideas are very fragile, so you have to be tender when they are in development. I realized that if he pissed on this, it would be so sad because I knew it was so important."

Just about everyone else was already sold. "It's just one of those things where instantly anyone who saw it was like, 'This is the coolest thing I've ever seen,'" Huppi recalls. "People's eyes would light up when they used it, when they saw this thing and played with it. And so we knew that there was something really magical about this."

The question was—would Steve Jobs think so too? After all, Jobs was the ultimate authority—he could kill the project with a word if he didn't see the potential.

The rig worked. The demos were compelling. They made it clear that instead of clicking and typing, you could touch, drag, toss, and manipulate information in a more fluid, intuitive way.

"Jony felt it was time to show it to Steve Jobs," Huppi says. At this point, it was as much a matter of timing as anything else. "If you caught Steve on a bad day, everything he saw was shit, and it was like, 'Don't ever show this to me again. Ever.' So you have to be very careful about reading him and knowing when to show him things."

In the meantime, the table-sized rig sat in the secret surveillance lab on Infinite Loop, projecting the shapes of the future onto a blank white sheet of paper.

CHAPTER 1

A Smarter Phone

Simon says, Show us the road to the smartphone

Stop me if you've heard this one before.

One day, a visionary innovator at one of the world's best-known technology companies decided that the future of communication lay in combining mobile phones with computing power. The key, he believed, was ensuring this new device functioned intuitively, so a user could pick it up and it would already feel familiar. It would have a touchscreen you could use to control the device with your fingers. It would have an easy-to-navigate home screen filled with icons you could tap to activate. It would have internet access and email. It would have games and apps.

First, though, a prototype had to be built in time to show it off to the world at a very public demonstration, where the eyes of the media would be fixed on him. In order to meet the deadline, the visionary pushed his team to the breaking point. Tensions mounted. Technology failed, then worked, then failed again. Miraculously, on the day of the big demo, the new hybrid phone was—barely—a go.

The innovator stepped into the spotlight and promised a phone that would change everything.

And so the smartphone was born.

The year was 1993.

The visionary inventor was Frank Canova Jr., who was working as an engineer in IBM's Boca Raton, Florida, labs. Canova conceived, patented, and prototyped what is widely agreed to be the first smartphone, the Simon Personal Communicator, in 1992. That was a year before the World Wide Web was opened to the public and a decade and a half before Steve Jobs debuted the iPhone.

While the iPhone was the first smartphone to go fully mainstream, it wasn't actually a breakthrough invention in its own right.

"I really don't see the iPhone as an invention so much as a compilation of technologies and a success in smart packaging," says Chris Garcia, the curator of the Computer History Museum, the world's largest collection of computer-related artifacts. "The iPhone is a confluence technology. It's not about innovation in any field," he says.

The most basic innovation of the smartphone was the introduction of a computer into a device that every household in the nation had access to: the telephone. The choices that were made to render the phone smart, like using a touchscreen interface and foregrounding apps, would have serious ramifications in shaping the modern world. And those foundations were set well over two decades ago.

As the renowned computer scientist Bill Buxton puts it, "The innovations of the Simon are reflected in virtually all modern touchscreen phones."

By 1994, Frank Canova had helped IBM not just invent but bring to market a smartphone that anticipated most of the core functions of the iPhone. The next generation of the Simon, the Neon, never made it to the market, but its screen rotated when you rotated the phone—a signature feature of the iPhone. Yet today, the Simon is merely a curious footnote in computing history. So the question is: Why didn't the Simon become the first iPhone?

"It's all about time frames," Canova says with a wry smile, holding up the first smartphone. It's black, boxy, and the size of a brick. "The technologies actually just barely allowed us to make this kind of phone."

We're sitting in his spacious, slightly cluttered office in Santa Clara, the heart of Silicon Valley, in the shadow of the amusement park Great America. Canova now works for Coherent, an industrial laser company, managing a team of engineers. It's a twenty-minute drive to Cupertino. And he's holding the third smartphone ever to roll off an assembly line: the Simon, serial number 3. He won't have it for much longer—historians are finally wising up to its value, and he's about to ship it off to the Smithsonian.

"It's a computer, so you have to boot it up," Canova says with a laugh as Simon emits a distinctly 1990s-flavored beep. Its yellow-green LCD screen lights up, and when I touch icons, like one named Address Book, they open new applications. The battery no longer lasts for more than a few seconds, so we keep it plugged in, but otherwise, it works seamlessly. The number pad is responsive. There's a slide-tile game. Sure enough, it feels like an 8-bit iPhone.

"So this is Simon the product," he says. "The year prior to this coming out, in 1992, we had a prototype. I wrote a bunch of apps for the technology demo—we had all sorts of, I'll call it visionary things, I wanted to put on there, so I included a map, GPS, stock quotes. We had an app for a whole bunch of things; we had games." Since the Cloud didn't exist yet, and large hard drives were too big to fit in a handset, many apps couldn't be included on the machine itself. The plan was to create a system to support add-on cards that would plug in to enable functionalities like GPS. He had proposed an old-school, IRL app store.

Canova, who is now in his fifties, is energetic and sharp. His head is clean-shaven and he sports a thick, graying mustache and a quick, mischievous smile. He grew up in Florida with a love of tinkering and gadgets; he was more Wozniak than Jobs, and experimented with hardware in his spare time. "I was a hacker, and hackers, well, from that era—hackers meant you could build a computer from scratch. So, I was building computers," he says, some based on the motherboard designs of Steve Wozniak. "It was unfortunate, I'll call it, to live in Florida, outside of where the [Silicon] Valley stuff was going on."

He graduated with a degree in electrical engineering from the Florida Institute of Technology and went to work for IBM. He stayed at the company for sixteen years, rising through the ranks thanks to his mastery of both hardware and software. In the 1980s, he joined an "advanced research team" that was in charge of engineering IBM's first laptop computer and making it as small as possible. "One of the goals there was just to make a computer that could fit in your shirt pocket," Canova says.

But the researchers didn't have the technology to make a computer that small. Then, as the laptop project was hitting a wall, one of IBM's neighbors presented the team with a fortuitous opportunity. "In Boca Raton, we had Motorola literally right down the street from us. Big plant, making all sorts of wireless products. They were very popular at the time," Canova says. That's putting it mildly: In the early 1990s, Motorola was the largest seller of cell phones in the world. And its Florida branch had an unusual business philosophy for the time—they were interested in sharing notes and collaborating with IBM. The engineers began exploring ways that they might combine the two tech giants' product lines. "Everyone was thinking, *How do you put a radio inside a desktop computer?*"

But Canova had grander ambitions. "It immediately became clear to me that, no, you don't want to have something to look like a computer at all. If you're making a radio, you want it to be portable. You want to hold it in one hand; you want to have it be intuitive, so what could your thumb do? You don't want to select something and have to type a command in every time you want it to do something, which in the DOS era is how you started programs." He didn't know what to call it, but Frank Canova wanted to build a smartphone.

"We approached Motorola about doing a joint project, essentially a smartphone project, and Motorola said no. They basically said, 'Well, we're not so sure about this shaky idea of yours,'" Canova says. But they agreed to support the team behind the scenes and provide Canova's group with the latest phone models. "We had to rub the names off of stuff, we had to paint over Motorola logos, because we were using their parts for this very first prototype of a smartphone back then."

Motorola wanted nothing to do with the first smartphone. Soon it became clear that IBM wasn't so sure either. "Honestly, IBM really wasn't interested in this business," Canova tells me. But he was convinced that he'd clawed at the kernel of something groundbreaking. He just needed the funding to prove it. He'd already convinced one of the sales managers to go to bat for the Simon, but that manager still had to win over his boss.

"The way he sold it," Canova says, "is I gave him this list of stuff you could do with a smartphone. So he took a big bag and he walked it over with all sorts of toys to our head guy who was running the Boca Raton site, and he said, 'Okay, we need funding. And it's funding for a thing that's going to do a lot of stuff.' So he pulled out a calculator and plopped it on his desk. Pulled out a GPS radio and plopped it on his desk. Pulled out a big book and maps, says, 'It's going to do that. And this.' And he starts filling up his table with all this stuff. And he says, 'You know, all this is going to be in one device. It's not going to be all this separate stuff.'"

The one-device presentation worked back in 1992 too. Frank's team got the funding and scrambled to build a functioning proto-type to show off at a technologies-of-the-future booth at COMDEX, then a major trade show. The team worked such long hours that Canova's newborn baby became a familiar sight in the lab—one of the only ways the new father could squeeze in time with him was to bring him to IBM.

The blitz paid off. The project briefly drew acclaim in the media, and IBM directed more resources to Canova's team. "To me, it was fundamental to make that interface as easy to use as a phone you could just pick up," Canova says. And that's exactly what IBM did.

"The Simon was ahead of its time in so many different ways," Canova says, a bit wistfully. That's an understatement. Smartphones wouldn't conquer the world for two more decades.

★ ★ ★

"There's really nothing new," Matt Novak says. "Apple and Samsung can believe that they invented these technologies, but there's always something that predated them, at least on paper."

Novak runs *Paleofuture,* a blog dedicated to collecting and analyz-ing the past's futuristic fantasies and predictions, and we're talking about the modern conception of the smartphone and the long his-tory of similar devices—both real and imagined—that preceded the iPhone, and even the Simon.

Visions of iPhone-like devices can be traced back to the late 1800s. One of the earliest and most striking is an 1879 cartoon by George du Maurier that appeared in the satirical *Punch Almanack*. Titled "Edi-son's Telephonoscope," it's a winking speculation about what it might look like if the famed American inventor managed to combine the telephone with a transmitter of moving images.

PUNCH'S ALMANACK FOR 1879.

EDISON'S TELEPHONOSCOPE (TRANSMITS LIGHT AS WELL AS SOUND).

The caption reads as follows:

(Every evening, before going to bed, Pater- and Materfamilias set up an electric camera-obscura over their bedroom mantel-piece, and gladden their eyes with the sight of their Children at the Antipodes, and converse gaily with them through the wire.)

Paterfamilias (in Wilton Place): "Beatrice, come closer, I want to whisper."

Beatrice (from Ceylon): "Yes, Papa dear."

Paterfamilias: "Who is that charming young Lady playing on Charlie's side?"

Beatrix: "She's just come over from England, Papa. I'll introduce you to her as soon as the Game's over!"

If you translate that Victorian vernacular into modern English and squint a little, you see wealthy parents FaceTiming with their kids away at summer camp. Maurier's speculation made the same promises that smartphone advertisers do today—the promise of never missing a moment with your friends and family, of unfettered communication, of having a portal into any part of the world.

In 1890, the futurist and satirist Albert Robida describes another telephonoscope in his illustrated novel *The Twentieth Century*. This one transmits both "dialogue and music" and the scene of a place itself "on a crystal disc with the clarity of direct visibility.... Thus we could—what a wonder (!)—become a witness in Paris of an event that took place a thousand miles away from Europe."

Many of these visions, I should note, were satirical—they saw the connected, electrified world as being full of absurdities and distractions—so the fact that the prophecies proved accurate shouldn't necessarily be cause for celebration.

Robida imagined people using his tScope (twenty-first-century branding was still beyond his grasp, so I've taken the liberty of helping him out) for entertainment—watching plays, sports, or news from afar; Maurier pictured people using it to stay ultraconnected with family and friends. Those are two of the most powerful draws of the smartphone today; two of its key functions—speed-of-light social networking and audiovisual communication—were outlined as early as the 1870s.

These ideas, whether fantastic or feasible, are constantly patched into what some academics term *technoculture*, the interplay between, well, technology and culture. It's a firmament of ideas that drives both invention and imagination. So it's not really surprising these particular smartphonic concepts, visions, and fantasies sprang up in the late 1800s. The electric revolution was fully under way then, driven by a flurry of already substantiated inventions, each stemming, mostly, from the telegraph.

* * *

The first optical telegraphs, or line-of-sight semaphores, were put into use during the French Revolution, to transmit military information between France and Austria. They could only transmit the equivalent of two words per minute, but information could suddenly travel many miles. Still, the general concept is ancient: Imagine the feeling of recognizing friendly code in a plume of smoke after spending the day warding off invaders of the Great Wall of China in 900 B.C.—it'd be at least as satisfying as getting a notification of a fresh round of Likes.

The telegraph took off in 1837, around when Samuel Morse commercialized the electrical variant, allowing data—via his eponymous code—to be carried across vast lengths of wire. "In a historical sense, the computer is no more than an instantaneous telegraph with a prodigious memory, and all the communications inventions in between have simply been elaborations on the telegraph's original work," according to the history of technology scholar Carolyn Marvin.

"In the long transformation that begins with the first application of electricity to communication, the last quarter of the nineteenth century has a special importance," Marvin writes. "Five proto–mass media of the twentieth century were invented during this period: the telephone, phonograph, electric light, wireless, and cinema." If you're counting, these are the prime ingredients for the smartphone you've got in your pocket right now.

A lot of technological progress has resulted from chasing those germs to their logical conclusions; high-res video, infinite playlists, LTE wireless networks, among others. But ultimately, of all the early transformative electric technologies, it's the phone that became the vessel for the rest.

* * *

"It's a phone first; it wasn't a computer at all," Canova says of his Simon. "It did have to have all of these features behind it which needed a computer, but you shouldn't expose the computer to the end user. You have to expose a very simple, basic user interface; you want the computers to be invisible."

At his office desk, he swings over to his landline phone, picks it up, and puts it to his ear. "And the phone's interface is easy and natural," he says. In the 1990s, everybody knew how to use it, because there are few devices more fundamental to modern civilization than the telephone.

A century before, however, the telephone was so novel that many investors and officials considered it a toy. Even so, Alexander Graham Bell wasn't the first to pioneer the concept. The idea of transmitting sound over an electric telegraph hung so thick in the air in the 1870s that some half a dozen figures are routinely placed in consideration for the phone's inventorship, including Elisha Gray, the electrical engineer who filed a similar patent on the very same day as Bell.

But Bell was a determined developer, presenter, and marketer, a lot like his contemporary Thomas Edison and a lot like Steve Jobs. He was also a gifted linguist and an educator who developed programs that helped the deaf learn to speak.

According to Bell, the telephone is said to have begun, like many myth-draped American inventions, with an epiphany. "If I could make a current of electricity vary in intensity precisely as the air varies in density during the production of sound," Bell said, "I should be able to transmit speech telegraphically." According to Herbert N. Casson's 1910 history of the telephone, Bell "dreamed of replacing the telegraph and its cumbrous sign-language by a new machine that would carry, not dots and dashes, but the human voice. 'If I can make a deaf-mute talk,' he said, 'I can make iron talk.'" Initially, he envisioned placing a harp at one end of a wire and a "speaking-trumpet" at the other; the tone of a voice spoken into the trumpet would then be reproduced by the harp strings. He was also testing new technologies to improve his Visible Speech program when he mentioned his experiments to a surgeon friend, Dr. Clarence J. Blake. "Why don't you use a real ear?" he asked. Bell was game.

The surgeon cut an ear from a dead man's head, including its eardrum and the associated bones. Bell took the skull fragment and arranged it so a straw touched the eardrum at one end and a piece of smoked glass at the other. When he spoke loudly into the ear, the vibrations of the eardrum made tiny markings on the glass.

"It was one of the most extraordinary incidents in the whole history of the telephone," Casson noted. "To an uninitiated onlooker, nothing could have been more ghastly or absurd. How could anyone have interpreted the gruesome joy of this young professor with the pale face and the black eyes, who stood earnestly singing, whispering, and shouting into a dead man's ear? What sort of a wizard must he be, or ghoul, or madman? And in Salem, too, the home of the witchcraft superstition! Certainly it would not have gone well with Bell had he lived two centuries earlier and been caught at such black magic." Through the experiment, Bell noticed that the thin eardrum could effectively transmit vibrations through bones. So he imagined a "membrane telephone"—two iron disks, à la eardrums, placed far apart and connected by an electrified wire. One would catch the vibrations of sound, the other would reproduce them—this was the theoretical basis for the telephone. Somehow, it's fitting that there's an actual human ear ingrained in the technical DNA of the phone. Bell won a patent for his telephone in 1876, and today it's widely considered one of the most valuable ever awarded.

After the technology was proven to work, Bell had a hell of a time trying to get anyone to think of it as much more than a scientific curiosity, though the effect of a voice transporting itself across an electrical wire was enough to turn heads and draw crowds. He took the invention to the Philadelphia Centennial, where he displayed it for amused audiences. Bell, an astute pitchman, hit the lecture circuit to show off his telephone and gave what were basically technology demos—early Steve Jobs–like keynotes. "Bell, in eloquent rhapsodies, painted word-pictures of a universal telephone," Casson wrote. By 1910, there were seven million telephones across the United States, population ninety-two million. "It is now in most places taken for granted, as though it were a part of the natural phenomena of this planet." It was the original phone that began our century-long drift toward being always connected, always available.

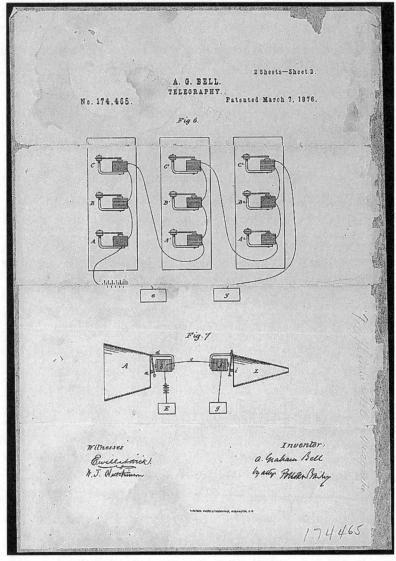

Graham Bell's infamous, "most valuable" patent, filed in 1876.

The next step was to cut the cord and make the telephone mobile—an idea that was in the air by the early 1900s. The satirical magazine *Punch* presciently ran a cartoon in its Forecasts for 1907 issue that depicted the future of mobile communications: A married couple sitting on the lawn, facing away from each other, engrossed in their devices. The caption reads: *These two figures are not communicating with one another. The lady is receiving an amatory message, and the gentleman some racing results.* That cartoon was lampooning the growing impact of telephones on society, satirizing a grim future where individuals sat alone next to one another, engrossed in the output of their devices and ignoring their immediate surroundings—lol?

DECEMBER 26, 1906.]   PUNCH, OR THE LONDON CHARIVARI.   451

FORECASTS FOR 1907.

IV.—DEVELOPMENT OF WIRELESS TELEGRAPHY. SCENE IN HYDE PARK.
[These two figures are not communicating with one another. The lady is receiving an amatory message, and the gentleman some racing results.]

The first truly mobile phone was, quite literally, a car phone. In 1910, the tinkerer and inventor Lars Magnus Ericsson built a telephone into his wife's car; he used a pole to run a wire up to the telephone lines that hung over the roads of rural Sweden. "Enough power for a telephone could be generated by cranking a handle, and, while Ericsson's mobile telephone was in a sense a mere toy, it did work," Jon Agar, the mobile-phone historian, notes. The company named after this invention, of course, would go on to become one of the biggest mobile companies in the world.

In 1917, a Finnish inventor named Eric Tigerstedt—whose groundbreaking work in acoustics and microphones earned him the nickname

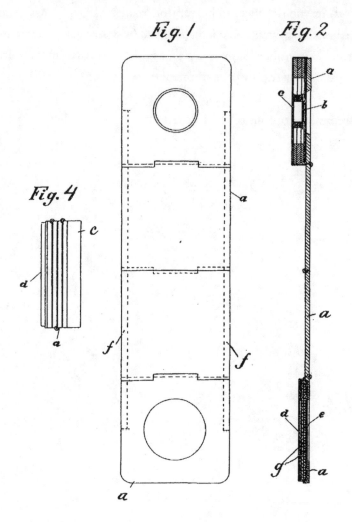

Eric Tigerstedt's "very thin" mobile-phone patent, circa 1917.

of "Thomas Edison of Finland"—successfully filed a patent for what appears to be the first truly mobile phone. In Danish patent no. 22901, Tigerstedt described his invention as a "pocket-size folding telephone with a very thin carbon microphone." It's more of a direct precursor to a flip phone, but it shares some distinct design and aesthetic features with the iPhone—thin, minimalist, compact. It's the earliest design for a mobile phone I've seen that feels truly modern. New ideas about handheld devices, networks, and data-sharing were beginning to emerge at that time as well—ideas presaging the internet, mobile computing, and global interconnectivity—at least from the better futurists of the day.

"When wireless is perfectly applied the whole earth will be converted into a huge brain, which in fact it is, all things being particles of a real and rhythmic whole," the famed scientist and inventor Nikola Tesla told *Collier's* magazine. "We shall be able to communicate with one another instantly, irrespective of distance. Not only this, but through television and telephony we shall see and hear one another as perfectly as though we were face to face, despite intervening distances of thousands of miles; and the instruments through which we shall be able to do his will be amazingly simple compared with our present telephone. A man will be able to carry one in his vest pocket."

His technological predictions were more on target than his sartorial ones—pocketed vests are out, obviously—but the outline of a smartphone-like technology, replete with the ability to connect to a globe-spanning, internet-like "brain," feels prescient.

Other keystones of the smartphone, like the touchscreen, were slipping into the technoculture too. The vision of a touchscreen-operated communications device that lets people interact or receive real-time information from around the world would become both a mainstay of science fiction and a pursuit of real-life engineering. Sometimes, it'd be hard to tell the difference.

In the 1940s and 1950s, some of the most influential computer scientists believed that personal computers would one day serve as knowledge augmenters—devices that would help people navigate an increasingly complex world. Vannevar Bush, a brilliant engineer and onetime head of the U.S. Office of Scientific Research and Development, envisioned the memex, a "memory index" device that would allow users to access vast libraries of data with the touch of a

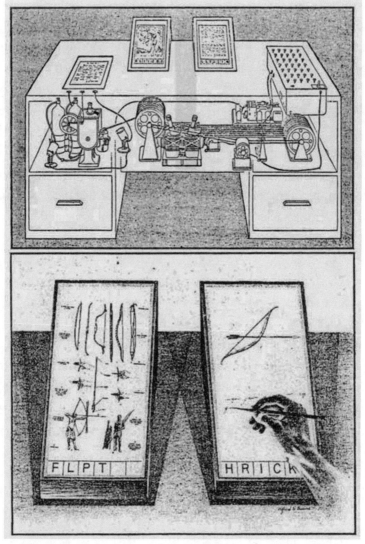

Vannevar Bush's memex, as diagrammed in Life *magazine in 1945.*

hand. His colleague and disciple J.C.R. Licklider, meanwhile, had foreseen the dawning age of human-computer symbiosis: "The hope is that, in not too many years, human brains and computing machines will be coupled together very tightly and that the resulting partnership will think as no human brain has ever thought," he wrote in 1950. Neither had any inkling that the vessel that would ultimately tightly couple brain to computer—that would enable that human-machine symbiosis—would be the cell phone.

In fact, the spark for modern computers and modern mobile phones was struck in the same space, foreshadowing the closeness with which they'd be bound. Cell phones became feasible pretty much immediately after the discovery of transistors, the key ingredient to modern computers, at Bell Labs.

Science fiction shaped what would become the smartphone as well, with two sources of inspiration looming large: "*Star Trek*. No doubt," Garcia says. "The tricorder and the communicator are direct influences, and I've spoken to several innovators who have specifically cited *Trek*." The second is *2001: A Space Odyssey,* which featured a device called a Newspad. "I think *2001* is the most mainstream representation of an iPhone- or iPad-size device in the late sixties," Novak says, "if you look at the Newspad in *2001,* I mean, that's an iPad."

Around the same time, Alan Kay designed the first mobile computer, the Dynabook: "A combination of this 'carry anywhere' device and a global information utility, such as ARPA network or two-way cable TV, will bring the libraries and schools (not to mention stores and billboards) of the world to the home."

Computers and cell phones would develop on separate tracks for the next half a century—researchers made smaller, faster, more multifunctional phones and computers until eventually they both were small enough to be smashed together.

The first device to be packaged explicitly as a "smartphone" was the Ericsson R380—a standard-looking cell phone that flipped open to reveal a touchscreen that could be used with a stylus. Nokia had phones that could run apps and play music. There was even another device, launched in 1998, that was actually called the iPhone—it was billed as a three-in-one "Internet Touchscreen Telephone" that functioned as an email reader, telephone, and internet terminal and was sold by a company called InfoGear.

"Without those, the iPhone would never have happened," Garcia says, "and I'll add another note—if any of those 1990s handhelds had succeeded, the iPhone never would have happened, because Apple would not have seen a field ready to be plucked!"

Get Smart

Frank Canova saw that field. But back in 1993, he was anxious.

"I walked outside, took a deep breath," he says, describing the day of Simon's first public demo. "I called the guys back in Florida and said, 'We're all set up, we're ready to go.' HQ was nervous. They had a backup—they didn't know if we were going to be ready." Canova grows excited as he recounts the story.

"There was a moment there, when I'm standing outside the convention center, when I had my calendar on the phone, and I could talk to a person and share with them what the schedule was. We even had it set up so they could send me a message and update my calendar from headquarters. From Florida. And there was that moment going, Wow, this is totally different. This isn't your IBM PC at that stage, this wasn't a classic desktop computer with a DOS prompt. This wasn't a cell phone, where you could make a verbal call. This was a way to interconnect people. And that was the point. That moment, where I stood outside of COMDEX, where I had a chance to take a deep breath and realize that this was about to change the world."

If this were a Hollywood movie, or even a TED Talk or a business-management bestseller, this is when all the hard work would pay off. This is when, having overcome the odds, the Simoneers, as they'd taken to calling themselves, would launch a bestselling, world-changing product and put it on retail shelves around the world.

It didn't happen.

IBM sold only fifty thousand Simons over the six months it was available, between 1994 and 1995, before the company discontinued the product. Yet when I told people that I was going to interview the man who held the first smartphone patent, the reaction was universal: He must be loaded. "Ha, well, as you can see, I'm not," Frank says, gesturing to his engineer's office—by no means spartan, but far from opulent. "IBM owns the patent anyway. I do get called to defend prior-art patents just about every year, to help companies show that smartphones go way back."

There are a number of reasons that the Simon didn't take off. (In business parlance, it *flopped* or *failed*, but those are misnomers, since it's hard to argue that a crucial iPhone forebear failed—you wouldn't say Einstein's grandfather failed because he didn't introduce the theory of relativity himself). Some of the reasons are obvious: It was expensive, retailing for $895. It was bulky, heavy, and, because this was before the mass adoption of Wi-Fi, it could be used to send email only via dial-up. And, unlike the iPhone, its media capabilities were incredibly limited; it couldn't play high-quality video or music, and its games were crude.

"And let's be honest, it's ugly as hell," Canova says with a laugh. But it had to be, to house the hardware. As Frank says, it's all about time frames.

Think of it this way: Steve Jobs is one of the most celebrated entrepreneurs in modern history. As I'm writing this, Frank Canova doesn't even have a Wikipedia page. (By the time this gets published, he very well could, of course, and it may have been written on a smartphone.) Most of the iPhone engineers I spoke with didn't cite the Simon as a major influence; some hadn't even heard of it, and some had forgotten about it. It's nonetheless undeniable that the two phones have a slew of overlapping functionalities and philosophies. There's something that seems almost universal about the devices, maybe because their inventors were drawing from a rich shared history of technological concepts and pop-culture predictions.

It's hard to shake the sense that the Simon was the iPhone in chrysalis, however obscured by black plastic and its now-comical size. The point isn't that Apple ripped off the Simon. It's that the conceptual framework for the smartphone, what people imagined they could do with a mobile computer, has been around far, far longer than the iPhone. Far longer than Simon, even.

"It's this push and pull," Novak says. "There's this 2012 Tim Cook interview with Brian Williams. Tim Cook holds up his iPhone and says, 'This is *The Jetsons*. I grew up watching *The Jetsons,* and this is *The Jetsons.*' Of course it's not. But it embodies what he thought, growing up, futuristic technology looked like . . . I spent 2012 looking up every episode in *The Jetsons,* and there is not a single device you could construe as an iPhone. But Cook's memory is such that it originated there. Every piece of future fiction is a Rorschach test."

The smartphone, like every other breakthrough technology, is built on the sweat, ideas, and inspiration of countless people. Technological progress is incremental, collective, and deeply rhizomatic, not spontaneous. "The evolution to the iPhone is really a multiverse," Garcia says. "No string of technologies leads to only one destination; each innovation leads to a series of new innovations."

The technologies that shape our lives rarely emerge suddenly and out of nowhere; they are part of an incomprehensibly lengthy, tangled, and fluid process brought about by contributors who are mostly invisible to us. It's a very long road back from the bleeding edge.

★ ★ ★

The story of the ideas and breakthroughs that eventually wound together into the smartphone stretches back over a century; the story of the raw stuff, the elemental materials that must come together to produce an actual *unit* of smartphone stretches across the globe. As long as we're investigating the early beginnings of the iPhone, let's examine the foundation of the physical object too.

CHAPTER 2

Minephones

Digging out the core elements of the iPhone

Cerro Rico towers over the old colonial city of Potosí, Bolivia, like a giant dusty pyramid. You can see the "rich hill" from miles away as you wind up the highway to the city gates. The landmark also goes by a nickname: "The Mountain That Eats Men." The mines that gave rise to both monikers have been running since the mid-1500s—that's when freshly arrived Spaniards began conscripting indigenous Quechua Indians to mine Rico.

The Mountain That Eats Men bankrolled the Spanish Empire for hundreds of years. In the sixteenth century, some 60 percent of the world's silver was pulled out of its depths. By the seventeenth century, the mining boom had turned Potosí into one of the biggest cities in the world; 160,000 people—local natives, African slaves, and Spanish settlers—lived here, making the industrial hub larger than London at the time. More would come, and the mountain would swallow many of them. Between four and eight million people are believed to have perished there from cave-ins, silicosis, freezing, or starvation.

"Cerro Rico stands today as the first and probably most important monument to capitalism and to the ensuing industrial revolution," writes the anthropologist Jack Weatherford. In fact, "Potosí was the first city of capitalism, for it supplied the primary ingredient of capitalism—money. Potosí made the money that irrevocably changed the economic complexion of the world." South America's first currency mint still stands in its downtown square.

Today, Cerro Rico has been carved out so thoroughly that geologists say the whole mountain might collapse, taking Potosí down with it. Yet around fifteen thousand miners—thousands of them children, some as young as six years old—still work in the mines, prying tin, lead, zinc, and a little silver from its increasingly thin walls. And there's a good chance some of that tin is inside your iPhone right now.

★ ★ ★

We didn't last half an hour down there.

Anyone with the stomach for it can glimpse the inside of this deadly mine, as enterprising Potosínos offer tours of the tunnels and shafts that make up the labyrinth under Cerro Rico. My friend and colleague (and translator) Jason Koebler arranged for us to take the plunge. Our guide, Maria, who also works as an elementary-school teacher, tells us that the tours go only to the "safe" parts. Yes, she said, many still die in the mines every year, but the last two, killed last week, were just kids who got drunk and got lost and froze to death. We shouldn't worry, she says. Sure.

The plan was to don hard hats, boots, protective ponchos, and headlamps and descend a mile or so into Rico. Before driving us to the entrance, Maria stops at Miner's Market, where we buy coca leaves and a potent 96 percent alcohol solution to give as gifts to any laborers we might encounter. Up top, the sun beats down hot but the air stings cold. Look out of the mine's opening, past a cluster of rusty mine carts, and the city of Potosí is splayed out in the distance.

I'm nervous. Even if tourists spelunk here each week, even if children work here every day, this slipshod mine tunnel is still terrifying. Potosí is the highest-altitude major city in the world, and we are even higher than the city, at about fifteen thousand feet. The air is thin, and my breathing is short. One look at the splintery wooden beams that hold open the narrowing, pitch-dark mine shaft we're about to walk down, one lungful of the sulfuric air, and my only impulse is to turn back.

Thousands of workers do this every day. Before they do, they bribe the devil. Did I not mention that the miners of Cerro Rico worship the devil? If not *the* devil, then *a* devil, El Tío. Near the entrance to most mines, there's an altar with an obscene-looking effigy of El Tío. Cigarette butts and coca leaves are crammed in his mouth, and beer cans lie at his feet; miners leave offerings as a bid for good luck. God may rule the heaven and earth, but the devil holds sway in the subterranean. Jason, Maria, and I light him three cigarettes and get ready for the deep.

★ ★ ★

Mining on Cerro Rico is a decentralized affair. The site is nominally owned by Bolivia's state-run mining company Comibol, but miners don't draw pay from the state; they work essentially as freelancers in loose-knit cooperatives. They gather the tin, silver, zinc, and lead ores and sell them to smelters and processors, who in turn sell to larger commodity buyers. This freelance model, combined with the fact that Bolivia is one of the poorest countries in South America, makes regulating work in the mines difficult.

That lack of oversight helps explain why as many as three thousand children are believed to work in Cerro Rico. A joint study conducted in 2005 by UNICEF, the National Institute of Statistics, and the International Labor Organization found seven thousand children working in mines in the Bolivian cities of Potosí, Oruro, and La Paz. According to 2009's *The World of Child Labor: An Historical and Regional Survey,* child labor was also found in mining centers across the region, including Huanuni and Antequera. So many children work in Bolivia that in 2014, the nation amended its child labor laws to allow ten-year-olds to do some work legally. That does not include mining—it's technically illegal for children of any age to work in the mines. But lack of enforcement and the cooperative structure make it easy for children to slip through the cracks. In 2008 alone, sixty children were killed in mining accidents at Cerro Rico. Maria tells us that the children work the deepest in the mines, in smaller, hard-to-reach places that are less picked over. It's high-risk, boom-or-bust work, and children will often follow their fathers into the mine to supplement the family income or pay for their own school supplies. Mining is one of the most profitable jobs an unskilled laborer can find, due in part to the steep risks.

Ifran Manene, an ex-miner who now works as a guide, started laboring here when he was thirteen. His father was a miner who spent his life working Cerro Rico. Ifran joined him as a teenager to help supplement the family income and worked alongside him for the next seven years. Today, Manene's father suffers from silicosis, a lung disease that afflicts many who spend years in the mine inhaling silica dust and other harmful chemicals—part of the reason why the life expectancy of a full-time miner in Cerro Rico is forty.

Workers get paid by the quantity of salable minerals they pry from Rico's walls, not by the hour. They use pickaxes and dynamite to break the rock free and load it into mine carts for transportation; the workers are said to distrust more efficient technologies because they would eliminate jobs. As a result, the mining inside Cerro Rico looks a lot like it did hundreds of years ago.

On a good day, these miners can make fifty dollars each, which is a hefty sum here. If they don't manage to find any significant amount of silver, tin, lead, or zinc, they make nothing. They sell the minerals to a local processor, who will smelt small quantities on-site and ship larger amounts of ore out of the city to an industrial-size smelter.

Silver and zinc are shipped to Chile by rail. Tin is shipped north to EM Vinto, Bolivia's state-run tin smelter, or to Operaciones Metalúrgicas S.A. (OMSA), a private one. And from there, that tin can make its way into Apple products.

"About half of all tin mined today goes to make the solder that binds the components inside our electronics," Bloomberg reported in 2014. Solder is made almost entirely of tin.

So, I think; metal mined by men and children wielding the most primitive of tools in one of the world's largest and oldest continuously running mines—the same mine that bankrolled the sixteenth century's richest empire—winds up inside one of today's most cutting-edge devices. Which bankrolls one of the world's richest companies.

★ ★ ★

How do we know that Apple uses tin from EM Vinto? Simple: Apple says it does.

Apple lists the smelters in its supply chain as part of the Supplier Responsibility reports it makes available to the public. Both EM Vinto and OMSA are on that list. And I was able to confirm through multiple sources—through miners on the ground as well as industry analysts—that tin from Potosí does in fact flow to EM Vinto.

Thanks to an obscure amendment to the 2010 Dodd-Frank financial-reform bill aimed at discouraging companies from using conflict minerals from the Democratic Republic of the Congo, public companies must disclose the source of the so-called 3TG metals (tin, tantalum, tungsten, and gold) found in their products.

Apple says that it set about mapping its supply chain in 2010. In 2014, the company began publishing lists of the confirmed smelters it uses and said it was working to rid its supply chain of smelters buying conflict minerals altogether. (As of 2016, Apple had become the first in the industry to get all the smelters in its supply chain to agree to regular audits.)

This is no small feat. Apple uses dozens of third-party suppliers to produce components found in devices like the iPhone, and all of those use their own third-party suppliers to provide yet more parts and raw materials. It makes for a vast web of companies, organizations, and actors; Apple directly purchases few of the raw materials that wind up in its products. That's true of many companies that manufacture smartphones, computers, or complex machinery— most rely on a tangled web of third-party suppliers to produce their stuff.

It means your iPhone begins with thousands of miners working in often brutal conditions on nearly every continent to dredge up the raw elements that make its components possible.

* * *

What are those raw materials exactly? What is the iPhone actually composed of at its most elemental level? To find out, I asked David Michaud, a mining consultant who runs 911 Metallurgist, to help me determine the chemical composition of the iPhone. To our knowledge, it's the first time such an analysis has been conducted. Here's how it worked.

I bought a brand-new iPhone 6 at the flagship Fifth Avenue Apple Store in Manhattan in June of 2016 and shipped it to Michaud. He sent it to a metallurgy lab, which performed the following tests:

First, they weighed the device; it's 129 grams, as Apple advertises. The iPhone was then set inside an impact machine used for pulverizing rock, where, in a contained environment, a 55-kilogram hammer was dropped on it from 1.1 meters above. The lithium-ion battery caught fire. The entire mass of the phone was then recovered and pulverized. "It surprised me how difficult it was to destroy," Michaud says. The materials were then extracted and analyzed.

From that process, the scientists were able to identify the elements that make up the iPhone.

"It's twenty-four percent aluminum," Michaud says. "You can see the outside case as being aluminum. You wouldn't think that the case weighs a quarter of the device....Aluminum is very light. It's cheap; it's a dollar a pound."

The iPhone is 3 percent tungsten, which is commonly mined in Congo and used in vibrators and on the screen's electrodes. Cobalt, a key part of the batteries, is mined in Congo too. Gold's the most valuable metal inside the device, and there isn't much of it.

"There were no precious metals detected in any major quantities, maybe a dollar or two," Michaud says. "Nickel is worth nine dollars a pound and there's two grams of it." It's used in the iPhone's microphone.

There's more arsenic in the iPhone than any of the precious metals, about 0.6 grams, though the concentration is too low to be toxic. The amount of gallium was a surprise. "It's the only metal that is liquid at room temperature," Michaud says. "It's a by-product. You have to mine coal to get gallium." The amount of lead, however, was not. "The world has tried very hard to get rid of lead, but it is difficult to do."

The oxygen, hydrogen, and carbon found are associated with different alloys used throughout the phone. Indium tin oxide, for instance, is used as a conductor for the touchscreen. Aluminum oxides are found in the casing, and silicon oxides are used in the microchip, the iPhone's brain. That's where the arsenic and gallium go too.

Silicon accounts for 6 percent of the phone, the microchips inside. The batteries are a lot more than that: They're made of lithium, cobalt, and aluminum.

iPhone 6, 16GB model

Element	Chemical Symbol	Percent of iPhone by weight	Grams used in iPhone	Average cost per gram	Value of element in iPhone
Aluminum	Al	**24.14**	31.14	$ 0.0018	$ 0.055
Arsenic	As	**0.00**	0.01	$ 0.0022	$ -
Gold	Au	**0.01**	0.014	$ 40.00	$ 0.56
Bismuth	Bi	**0.02**	0.02	$ 0.0110	$ 0.0002
Carbon	C	**15.39**	19.85	$ 0.0022	$ -
Calcium	Ca	**0.34**	0.44	$ 0.0044	$ 0.002
Chlorine	Cl	**0.01**	0.01	$ 0.0011	$ -
Cobalt	Co	**5.11**	6.59	$ 0.0396	$ 0.261
Chrome	Cr	**3.83**	4.94	$ 0.0020	$ 0.010
Copper	Cu	**6.08**	7.84	$ 0.0059	$ 0.047
Iron	Fe	**14.44**	18.63	$ 0.0001	$ 0.002
Gallium	Ga	**0.01**	0.01	$ 0.3304	$ 0.003
Hydrogen	H	**4.28**	5.52	$ -	$ -
Potassium	K	**0.25**	0.33	$ 0.0003	$ -
Lithium	Li	**0.67**	0.87	$ 0.0198	$ 0.017
Magnesium	Mg	**0.51**	0.65	$ 0.0099	$ 0.006

Manganese	Mn	**0.23**	0.29	$	0.0077	$	0.002
Molybdenum	Mo	**0.02**	0.02	$	0.0176	$	0.000
Nickel	Ni	**2.10**	2.72	$	0.0099	$	0.027
Oxygen	O	**14.50**	18.71	$	-	$	-
Phosphorus	P	**0.03**	0.03	$	0.0001	$	-
Lead	Pb	**0.03**	0.04	$	0.0020	$	-
Sulfur	S	**0.34**	0.44	$	0.0001	$	-
Silicon	Si	**6.31**	8.14	$	0.0001	$	0.001
Tin	Sn	**0.51**	0.66	$	0.0198	$	0.013
Tantalum	Ta	**0.02**	0.02	$	0.1322	$	0.003
Titanium	Ti	**0.23**	0.30	$	0.0198	$	0.006
Tungsten	W	**0.02**	0.02	$	0.2203	$	0.004
Vanadium	V	**0.03**	0.04	$	0.0991	$	0.004
Zinc	Zn	**0.54**	0.69	$	0.0028	$	0.002
	TOTAL	**100%**	**129 grams**			**$**	**1.03**

And there are elements locked inside each iPhone that constitute too small a percentage of the phone's mass to show up in the analysis. In addition to precious metals like silver, there are crucial elements known as rare earth metals, like yttrium, neodymium, and cerium.

All of these elements, precious or abundant, have to be pulled out of the earth before they can be mixed into alloys, molded into compounds, or melted into the plastics that make up the iPhone. Apple does not disclose where its nonconflict minerals come from, but many sources have been reported over the years; here's a quick sampling of how some of the crucial elements in the iPhone are mined.

Aluminum

Aluminum is the most abundant metal on Earth. It's also the most abundant metal in your iPhone, due to its anodized casing. Aluminum comes from bauxite, which is often strip-mined, an operation that can devastate the natural landscape and imperil natural habitats. And it takes four tons of bauxite to produce one ton of aluminum, creating a load of excess waste. Aluminum smelters suck down a full 3.5 percent of the globe's power. In the process, they release greenhouse gases that are 9,200 times more potent than carbon dioxide.

Cobalt

Most of the cobalt that ends up in the iPhone is in its lithium-ion battery, and it comes from the Democratic Republic of the Congo. In 2016, the *Washington Post* found laborers working around the clock with hand tools in small-scale pits in DRC cobalt mines. They rarely wore protective gear, and the mines were almost totally unregulated. Child laborers toiled here too. "Deaths and injuries are common," the investigation found.

Tantalum

Around the time that Apple announced it had reaped the largest corporate profits of any public company in history, it verified that its tantalum suppliers were conflict-free. Tantalum was, for a long time, sourced largely from the DRC, where rebels and the army alike forced children and slaves to work in mines and used the mining profits to sustain their campaigns of violence. Mass rapes, child soldiers, and genocides have been bankrolled by the 3TG.

Rare Earth Metals

The iPhone's hundreds of components require a suite of rare earth metals — such as cerium, which is used in a solvent to polish touchscreens and to color glass, and neodymium, which makes powerful, tiny magnets and shows up in a lot of consumer electronic parts — and mining these elements is a complex, sometimes toxic affair.

Most rare earth metals come from a single place: Inner Mongolia, a semiautonomous zone in northern China. There, the by-products from mining have created a lake that's so gray, so drenched in toxic waste, that it's been dubbed "the worst place on earth" by the BBC. "Our lust for iPhones, flat-screen televisions, and the like created this lake," BBC investigator Tim Maughan, one of the few journalists to actually see the lake, told me.

Rare earths aren't *rare* in the way we typically interpret that term. They're not scarce; workers simply have to mine an awful lot of earth to get a small amount of, say, neodymium, which makes for an energy- and resource-intensive process and results in a lot of waste. Apple—and just about everyone else—outsources the operation to China, largely because the country doesn't have the environmental regulations that other nations do. (A U.S. company called Molycorp tried to mine cleanly for rare earth metals in the southwestern desert; it went bankrupt in 2014.) The BBC investigation revealed that the lake isn't just toxic, it's radioactive—clay collected from its bed tested at three times higher than the background radiation.

Tin

Bangka Island, in Indonesia, is home to about half of the tin smelters on Apple's list. This is probably because, as a *Bloomberg BusinessWeek* report revealed, it's where their manufacturing partner Foxconn sources its tin. The Bangka Island mines are chaotic and lethal. Miners flock to thousands of small-scale pits, each fifteen to forty feet deep, many of them illegal, and dig the tin out of the ground with pickaxes or by hand. The mine bosses frequently use tractors to create pits that leave almost vertical and unstable walls of earth that are apt to come crashing down on the laborers. In 2014, the fatality rate was one miner a week. After *Bloomberg* published that report, Apple sent an envoy to Indonesia and pledged to work with local groups and the environmental organization Friends of the Earth, though it's not entirely clear what the impact has been. Meanwhile, the mining operations have razed large parts of the island's flora, and miners have taken to dredging the seabed for ore, plowing through the reefs and aquatic habitats.

* * *

Michaud crunched the numbers to generate an estimate of how much earth had to be mined to create a single iPhone. Based on data provided by mining operations around the world, he determined that approximately 34 kilograms (75 pounds) of ore would have to be mined to produce the metals that make up a 129-gram iPhone. The raw metals in the whole thing are worth about one dollar total, and 56 percent of that value is the tiny amount of gold inside. Meanwhile, 92 percent of the rock mined yields metals that make up just 5 percent of the device's weight. It takes a lot of mining—and refining—to get small amounts of the iPhone's rarer trace elements, in other words.

A billion iPhones had been sold by 2016, which translates into 34 billion kilos (37 million tons) of mined rock. That's a lot of moved earth—and it leaves a mark. Each ton of ore processed for metal extraction requires around three tons of water. This means that each iPhone "polluted" around 100 liters (or 26 gallons) of water, Michaud tells me. Producing 1 billion iPhones has fouled 100 billion liters (or 26 billion gallons) of water.

Furthermore, extracting gold from a ton of ore typically requires about two and a half pounds (1,136 grams) of cyanide, Michaud says, as the chemical is used to dissolve and separate rock from precious metals. Because up to 18 of the 34 kilos of ore mined to produce each iPhone are mined in pursuit of gold, it would require 20.5 grams of cyanide to free enough gold to produce an iPhone.

So, according to Michaud's calculations, producing a single iPhone requires mining 34 kilos of ore, 100 liters of water, and 20.5 grams of cyanide, per industry average.

"That's what's shocking!" he says.

* * *

Deep in a mine shaft at Cerro Rico, Marie, Jason, and I are ducking under collapsed support beams, checking out the mineral deposits in the rock seams, shining our headlamps down forks in the tunnel that look like they might never end. It's deep-space black down here. Jason and I are both pretty tall and lanky. For stretches, the tunnel is only four feet high, forcing us to squat and waddle. The walls close in tight, and the air feels thick. Jason starts to get anxious. So I start to get anxious. Our guide takes the top off the bottle of moonshine we've brought as a gift for the miners and holds the bottle under our noses. It is indeed pretty efficient at delivering an ugly wake-up jolt.

A second later, my head hits the ceiling and a spill of sediment dusts my face. I'm taking video and blurry flash photos on my iPhone. The deposits on the walls — sulfur, maybe — are oddly beautiful.

Jason looks pale. I get it. The whole mountain is a ticking geological time bomb. It feels stupid to have this fear after a brief jog into a tunnel where thousands of people work every day, but here we are. Still, I bet most iPhone owners would start to lose it if they had to spend more than twenty minutes down here. Jason wants to turn back.

Before I know it, we're winding back through the dark, and finally that little circle of light comes around a corner.

Like I said: we didn't last half an hour down there.

* * *

Ifran Manene, the teenage miner turned tour guide, puts it bluntly. Two of his friends are in the hospital right now. His father is sick. "Every year, we have more than fifteen miners die" in Cerro Rico alone, he says. And he tells me this without a trace of lament, like it's perfect normal. The sum of this human cost is difficult to comprehend, and there are stories like this taking place on almost every continent behind many of the dozens of elements in the iPhone.

It's an uncomfortable fact, but one we'd do well to internalize: Miners working with primitive tools in deadly environments produce the feedstock for our devices. Many of the iPhone's base elements are dug out in conditions that most iPhone users wouldn't tolerate for even a few minutes. Cash-poor but resource-rich countries will face an uphill struggle as long as there's a desire for these metals—demand will continue to drive mining companies and commodities brokers to find ways to get them. These nations' governments, like Bolivia's, will struggle to regulate the industry. For the foreseeable future, miners will continue to do backbreaking, lung-infecting labor to bring us the ingredients of the iPhone.

* * *

There's another critical material we haven't discussed, and it's the first thing you touch after you grab your iPhone—its chemically strengthened, scrape resistant glass.

CHAPTER 3

Scratchproof

Break out the Gorilla Glass

It's a universal, gut-churning feeling—the phone slides from your hand, *just* out of range of your frantic lunge to catch it, and lands with a painful crack on the floor. Then, the swelling anxiety as you pick it up—you almost can't bear to look—to see if your screen survived. The sigh of relief when, amazingly, it did. Or, the sinking despair when it didn't. Still, when you consider the amount of abuse your phone endures—it barely registers a scratch after sharing front pocket real estate with your keys, being slid face-down across rough surfaces, or taking tumbles off tables and desks—the glass that coats its display is pretty remarkable. So is where it comes from.

If your grandparents ever served you a casserole in a white, indestructible-looking dish with blue cornflowers on the side, then you've eaten out of the material that would give rise to the glass that protects your iPhone. That dish is made of CorningWare, a ceramic-glass hybrid created by one of the nation's largest, oldest, and most inventive glass companies.

In the early 1950s, one of Corning's inventors, a chemist named Don Stookey, was experimenting with photosensitive glass in his lab at the company's headquarters in upstate New York. He placed a sample of lithium silicate into a furnace and set it to 600°C (about

1,100°F)—roughly the temperature of a pizza oven. Alas, a faulty controller allowed the temperature to climb to 900°C (about 1,650°F)—roughly the temperature of intermediate lava as it exits the earth. When Stookey realized this, he opened the furnace door expecting that both the experiment and his equipment would be ruined. To his surprise, he found the silicate had been transfigured into an off-white-colored plate. He tried to pull it out of the furnace, but it slipped out of his tongs and fell to the floor. Weirdly, it didn't break—it bounced.

Inventors had been stumbling around shatterproof glass for at least a half a century by then. In 1909, a French chemist and art deco artist named Édouard Bénédictus was climbing a ladder in his laboratory when he accidentally knocked a glass flask off the shelf. Instead of shattering and sending shards of glass flying, the flask cracked but stayed in one piece. Perplexed, Bénédictus studied the glass and realized it had contained cellulose nitrate, a liquid plastic, which had evaporated and left a thin film inside. That film had snagged the glass shards and prevented them from scattering on impact.

The artist and inventor spent the next twenty-four hours in a frenzy of experimentation; he knew that nascent automobile windshields were dangerously prone to shattering, and he saw a solution. Later that year, Bénédictus filed the world's first patent for shatterproof safety glass. But carmakers weren't initially interested in a more expensive glass, even if it was safer. It wasn't until World War I, when a version of Bénédictus's invention was used in the eyepieces of American soldiers' gas masks, that safety glass became cheap to manufacture. (Military-scale industrialization tends to have that effect.) And in 1919, a full decade after Bénédictus's happy accident, Henry Ford began incorporating the glass into his windshields.

But it was Don Stookey who invented the first synthetic glass-ceramic. Corning would go on to call it Pyroceram (it was the brink of the 1960s, and awkward, quasi-futuristic portmanteaus were all the rage). The stuff was light, harder than steel, and much, much stronger than typical glass. Corning sold it to the military, where it was used in missile nose cones. But the real boon came when Corning found synergy with another ascendant technology: the microwave. Corning's line of serving dishes—CorningWare—worked well in the futuristic food cooker. They sold like radiated hotcakes.

In the late 1950s, according to a famous bit of company lore, Corning's president, Bill Decker, had a chat with William Armistead, the company's chief of research and development. "Glass breaks," Decker remarked. "Why don't you fix that?"

CorningWare didn't break, but it was opaque. Given the material's success, the company effectively doubled its research and development budget. And thus, Corning launched its magnificently titled Project Muscle with the goal of creating still stronger, transparent glass. Its research team investigated all forms of glass strengthening that were known at the time, which mostly fell into two categories: the age-old technique of tempering, or strengthening glass with heat, and the newer one of layering types of glass that expanded at different rates when exposed to heat. When those various layers of glass cooled, the researchers hoped, they'd compress and strengthen the final product. The Project Muscle experiments, which hit full throttle in 1960 and 1961, combined both tempering and layering. It soon led to a new, ultrastrong, remarkably shatterproof—and scratchproof—glass.

"A breakthrough came when company scientists tweaked a recently developed method of reinforcing glass that involved dousing it in a bath of hot potassium salt," explains Bryan Gardiner, a reporter who investigated Corning's relationship with Apple in 2012. "They discovered that adding aluminum oxide to a given glass composition before the dip would result in remarkable strength and durability."

The ingenious chemical-strengthening process relied on a new method called ion exchange. First, sand—the core ingredient of most glass—is blended with chemicals to produce a sodium-heavy aluminosilicate. Then the glass is bathed in potassium salt and heated to 400°C (752°F). Because potassium is heavier than the sodium in the original blend, the "large ions are 'stuffed' into the glass surface, creating a state of compression," according to Corning. They called the new glass Chemcor. It was much, much stronger than ordinary glass. And you could still see through it.

Chemcor was fifteen times stronger than regular glass—it was said that the stuff could withstand pressures of up to 100,000 pounds per square inch. Of course, the researchers had to be sure, so they set out to stress-test the wonder glass. They hurled tumblers made of Chemcor glass off the roof of the research center onto a steel plate. Didn't break. So they stepped it up a bit; in the experiments, they chucked frozen chickens onto sheets of the new glass. Fortunately, Chemcor glass proved frozen-poultry-proof too.

By 1962, Corning figured the glass was ready for prime time. But Corning had no idea how to market Chemcor—or, rather, it had too many ideas. So Corning set up a press conference in downtown Manhattan to show it off and let the market come to it. They banged it, bent it, and twisted it, but failed to break it. The stunt generated good PR; thousands of inquiries about the glass poured in. Bell Telephone considered using Chemcor to vandal-proof its phone booths. Eyeglass makers had a look. And Corning itself developed some seventy ideas for potential product uses, including sturdy windows for jails and, yes, shatterproof windshields.

But, as with Bénédictus, the interest led to few takers. For automakers, who by then were using our French friend's laminated technique, Chemcor was simply *too* strong. When carmakers orchestrated crash tests, well—"Skulls were not left intact after colliding with it," Gardiner says. Windshields need to break in car accidents if the humans inside are to survive. Chemcor ended up in some of AMC's classic Javelin cars, but production was soon discontinued.

Forty-two million dollars had been invested in the product by 1969, and Chemcor was ready to strengthen the world's panes. But the market had spoken; nobody really wanted superstrong, costlier glass. It was just too expensive, too unique. Chemcor and Project Muscle were scrapped in 1971.

* * *

Three and a half decades later, in September 2006, just four months before Steve Jobs planned to reveal the iPhone to the world, he showed up at Apple HQ in a huff.

"Look at this," he said to a midlevel executive, holding up a prototype iPhone with scratch marks all over its plastic display—a victim of sharing his pocket with his keys. "Look at this. What's with the screen?"

"Well, Steve," the exec said, "we have a glass prototype, but it fails the one-meter drop test one hundred out of one hundred times—"

Jobs cut him off. "I just want to know if you are going make the fucking thing work."

That exchange may be notable for its snapshot of ultra-Jobs-ness, but it had real ramifications.

"We switched from plastic to glass at the very last minute, which was a curveball," Tony Fadell, the head of the original iPhone's engineering team, tells me with a laugh. "There were just so many things like that."

The original plan had been to ship the iPhone with a hard plexiglass display, as Apple had done with its iPod. Jobs's about-face gave the iPhone team less than a year to find a replacement that would pass that drop test. The problem was, there was nothing on the consumer-glass market that fit the bill—the glass on offer was mostly either too fragile and shatter-prone or too thick and unsexy. So, first, Apple tried to do its own in-house glass strengthening. The record is murky as to how long or how seriously this was tried— Apple didn't exactly have a massive materials-science division in the mid-2000s—but it was abandoned.

A friend of Jobs suggested he reach out to a man named Wendell Weeks, the CEO of a New York glass company called Corning. Long after inventing microwavable ceramic, Corning had continued to innovate—beyond Pyroceram, its researchers also invented low-loss optical fibers, in 1970, which helped wire the internet itself. In 2005, as stylish flip phones like the Razr were ascendant, Corning returned to the scrapped Chemcor effort to see if there might be a way to provide strong, affordable, scratchproof glass to cell phones. They code-named the project Gorilla Glass after the "toughness and beauty" of the iconic primate.

So when the Apple chief went to visit the head of Corning in its upstate New York headquarters, Weeks had a recently reinvigorated half-century-old research effort in full swing. Jobs told Weeks what they were looking for, and Weeks told him about Gorilla Glass.

The now-infamous exchange was well documented by Walter Isaacson in his biography *Steve Jobs*: Jobs told Weeks he doubted Gorilla Glass was good enough, and began explaining to the CEO of the nation's top glass company how glass was made. "Can you shut up," Weeks interrupted him, "and let me teach you some science?" It was one of the rare occasions Jobs was legitimately taken aback in a meeting like that, and he fell silent. Weeks took to the whiteboard instead, and outlined what made his glass superior. Jobs was sold, and, recovering his Jobsian flair, ordered as much as Corning could make — in a matter of months.

"We don't have the capacity," Weeks replied. "None of our plants make the glass now." He protested that it would be impossible to get the order scaled up in time.

"Don't be afraid," Jobs replied. "Get your mind around it. You can do it." According to Isaacson, Weeks shook his head in astonishment as he recounted the story. "We did it in under six months," he said. "We produced a glass that had never been made."

Corning had prototyped the stuff fifty years ago but never produced the material in any significant quantity. Within years, it'd be covering the face of nearly every smartphone on the market.

Gorilla Glass is made with a process called fusion draw. As Corning explains it, "molten glass is fed into a trough called 'an isopipe,' overfilling until the glass flows evenly over both sides. It then rejoins, or fuses, at the bottom, where it is drawn down to form a continuous sheet of flat glass that is so thin it is measured in microns." It's about the width of a sheet of aluminum foil. Next, robotic arms help smooth the overflow, and it's moved into the potassium baths and the ion exchange that gives it its strength.

Corning's Gorilla Glass is forged in a factory nestled between the rolling tobacco fields and sprawling cattle ranches of Harrodsburg, Kentucky (population 8,000). The plant employs hundreds of union workers and around a hundred engineers.

"The reason that a place like Corning comes to this area is to hire guys that have grown up on farms," Zach Ipson, a local farmer, told NPR in 2013. "They know how to work." Outside the idyllic town known for its bountiful tobacco harvests, a key component of one of the world's bestselling devices is forged in a state-of-the-art glass factory. It's one of the few parts of the iPhone that's manufactured in the United States. "When [I] tell someone where I live and where I work, they're surprised that we have this high-tech manufacturing operation in the bluegrass area that's known for bourbon and horses and farmland," engineer Shawn Marcum said.

Gorilla Glass is now one of the most important materials to the consumer electronics industry. It covers our phones and our tablets, and soon it may cover just about everything else. Corning has big plans; it imagines smart screens—made with its Gorilla Glass, of course—covering every surface of the increasingly smart home. Gorilla Glass may finally come to auto windshields, fifty years after Chemcor's initial market failure.

Securing that Apple contract helped the company thrive, and not only because the iPhone itself proved a hit. Samsung, Motorola, LG, and just about every other handset maker that rushed into the smartphone game in the wake of the iPhone's success turned to Corning.

The iPhone helped awaken the technology, but Project Muscle was already there, after waiting for decades in a shuttered research lab, to scratchproof the modern world.

A world that runs, increasingly, on touchscreens.

CHAPTER 4

Multitouched

How the iPhone became hands-on

The world's largest particle physics laboratory sprawls across the Franco-Swiss border like a hastily developed suburb. The sheer size of the labyrinthine office parks and stark buildings that make up the European Organization for Nuclear Research, better known as CERN, is overwhelming, even to those who work there.

"I still get lost here," says David Mazur, a legal expert with CERN's knowledge-transfer team and a member of our misfit tour group for the day, along with yours truly, a CERN spokesperson, and an engineer named Bent Stumpe. We take a couple wrong turns as we wander through an endless series of hallways. "There is no logic to the numbering of the buildings," Mazur says. We're currently in building 1, but the structure next door is building 50. "So someone finally made an app for iPhone to help people find their way. I use it all the time."

CERN is best known for its Large Hadron Collider, the particle accelerator that runs under the premises in a seventeen-mile subterranean ring. It's the facility where scientists found the Higgs boson, the so-called God particle. For decades, CERN has been host to a twenty-plus-nation collaboration, a haven that transcends geopolitical tensions to foster collaborative research. Major advances in our understanding of the nature of the universe have been made here. Almost as a by-product, so have major advances in more mundane areas, like engineering and computing.

We're shuffling up and down staircases, nodding greetings at students and academics, and gawking at Nobel-winning physicists. In one stairwell, we pass ninety-five-year-old Jack Steinberger, who won the prize in 1988 for discovering the muon neutrino. He still drops by all the time, Mazur says. We're pleasantly lost, looking for the birthplace of a piece of technology that history has largely forgotten: a touchscreen built in the early 1970s that was capable, its inventor says, of multitouch.

Multitouch, of course, is what Apple's ENRI team seized on when they were looking for a way to rewrite the language of how we talk to computers.

"We have invented a new technology called multitouch, which is phenomenal," Steve Jobs declared in the keynote announcing the iPhone. "It works like magic. You don't need a stylus. It's far more accurate than any touch display that's ever been shipped. It ignores unintended touches; it's super smart. You can do multi-finger gestures on it. And, boy, have we patented it." The crowd went wild.

But could it possibly be true?

It's clear why Jobs would want to lay claim to multitouch so aggressively: it set the iPhone a world apart from its competition. But if you define *multitouch* as a surface capable of detecting at least two or more simultaneous touches, the technology had existed, in various forms, for decades before the iPhone debuted. Much of it's history, however, remains obscured, its innovators forgotten or unrecognized.

Which brings us to Bent Stumpe. The Danish engineer built a touchscreen back in the 1970s to manage the control center for CERN's amazingly named Super Proton Synchrotron particle accelerator. He offered to take me on a tour of CERN, to show me "the places where the capacitive multitouch screen was born." See, Stumpe believes that there's a direct lineage from his touchscreen to the iPhone. It's "similar to identical" to the extent, he says, that Apple's patents may be invalid for failing to cite his system.

"The very first development was done in 1972 for use in the SPS accelerator and the principle was published in a CERN publication in 1973," he told me. "Already this screen was a true capacitive transparent multitouch screen."

So it came to pass that Stumpe picked me up from an Airbnb in Geneva one autumn morning. He's a spry seventy-eight; he has short white hair, and his expression defaults to a mischievous smile. His eyes broadcast a curious glint (Frank Canova had it too; let's call it the unrequited-inventor's spark). As we drove to CERN, he made amiable small talk and pointed out the landmarks.

There was a giant brutalist-looking dome, the Globe of Science and Innovation, and a fifteen-ton steel ribbon sculpture called *Wandering the Immeasurable*, which is also a pretty good way to describe the rest of the day.

Before we get to Stumpe's touchscreen, we stop by a site that was instrumental to the age of mobile computing, and modern computing, period—the birthplace of the World Wide Web. There would be no great desire for an "internet communicator" without it, after all.

★ ★ ★

Ground zero for the web, is, well, a pretty unremarkable office space. Apart from a commemorative plaque, it looks exactly the way you'd expect an office at a research center to look: functional, kind of drab. The future isn't made in crystal palaces, folks. But it was developed here, in the 1980s, when Tim Berners-Lee built what he'd taken to calling the World Wide Web. While trying to streamline the sharing of data between CERN's myriad physicists, he devised a system that linked pages of information together with hypertext.

That story is firmly planted in the annals of technology. Bent Stumpe's much lesser known step in the evolution of modern computing unfolded a stone's throw away, in a wooden hut within shouting distance of Berners-Lee's nook. Yes, one of the earliest multitouch-capable devices was developed in the same environment—same insti-tution, same setting—that the World Wide Web was born into, albeit a decade earlier. A major leap of the iPhone was that it used multitouch to allow us to interact with the web's bounty in a smooth, satisfying way. Yet there's no plaque for a touchscreen—it's just as invisible here as everywhere else. Stumpe's screen is a footnote that even technology historians have to squint to see.

Then again, most touchscreen innovators remain footnotes. It's a vital, underappreciated field, as ideas from remarkably disparate industries and disciplines had to flow together to bring multitouch to life. Some of the earliest touch-technology pioneers were musicians looking for ways to translate creative ideas into sounds. Others were technicians seeking more efficient ways to navigate data streams. An early tech "visionary" felt touch was the key to digital education. A later one felt it'd be healthier for people's hands than keyboards. Over the course of half a century, impassioned efforts to improve creativity, efficiency, education, and ergonomics combined to push touch and, eventually, *multi*touch into the iPhone, and into the mainstream.

* * *

In the wake of Steve Jobs's 2007 keynote, in which he mentioned that he and Apple had invented multitouch, Bill Buxton's in-box started filling up. "Can that be right?" "Didn't you do something like that years ago?"

If there's a generally recognized godfather of multitouch, it's prob-ably Buxton, whose research helped put him at the forefront of inter-action design. Buxton worked at the famed Xerox PARC in Silicon Valley and experimented with music technology with Bob Moog, and in 1984, his team developed a tablet-style device that allowed for continuous, multitouch sensing. "A Multi-Touch Three Dimensional Touch-Sensitive Tablet," a paper he co-authored at the University of Toronto in 1985, contains one of the first uses of the term.

Instead of answering each query that showed up in his email individually, Buxton compiled the answers to all of them into a single document and put it online. "Multitouch technologies have a long history," Buxton explains. "To put it in perspective, my group at the University of Toronto was working on multitouch in 1984, the same year that the first Macintosh computer was released, and we were not the first."

Who was, then? "Bob Boie, at Bell Labs, probably came up with the first working multitouch system that I ever saw," he tells me, "and almost nobody knows it. He never patented it." Like so many inventions, its parent company couldn't quite decide what to do with it.

Before we get to multitouch prototypes, though, Buxton says, if we really want to understand the root of touch technology, we need to look at electronic music.

"Musicians have a longer history of expressing powerful creative ideas through a technological intermediary than perhaps any other profession that ever has existed," Buxton says. "Some people would argue weapons, but they are perhaps less creative." Remember Elisha Gray, one of Graham Bell's prime telephone competitors? He's seen as a father of the synth. That was at the dawn of the twentieth century. "The history of the synthesizer goes way back," Buxton says, "and it goes way back in all different directions and it's really hard to say who invented what." There were different techniques used, he says, varying volume, pressure, or capacitance. "This is equally true in touchscreens," he adds.

"It is certainly true that a touch from the sense of a human perspective—like what humans are doing with their fingers—was always part of a musical instrument. Like how you hit a note, how you do the vibrato with a violin string and so on," Buxton says. "People started to work on circuits that were capable of capturing that kind of nuance. It wasn't just, 'Did I touch it or not?' but 'How hard did I touch it?' and 'If I move my fingers and so on, could it start to get louder?'"

One of the first to experiment with electronic, gesture-based music was Léon Thérémin. The Russian émigré's instrument—the theremin, clearly—was patented in 1928 and consisted of two antennas; one controlled pitch, the other loudness. It's a difficult instrument to play, and you probably know it best as the generator of retro-spooky sound effects in old sci-fi films and psychedelic rock tunes. But in its day, it was taken quite seriously, at least when it was in the hands of its star player, the virtuosa Clara Rockmore, who recorded duets with world-class musicians like Sergey Rachmaninoff.

The theremin inspired Robert Moog, who would go on to create pop music's most famous synthesizer. In addition to establishing a benchmark for how machines could interpret nuance when touched by human hands, he laid out a form for touchpads. "At the same time, Bob also started making touch-sensitive touchpads to driver synthesizers," Buxton says. Of course, he wasn't necessarily the *first*—one of his peers, the Canadian academic Hugh Le Caine, made capacitive-touch sensors. (Recall, that's the more complex kind of touchscreen that works by sensing when a human finger creates a change in capacitance.) Then there was Don Buchla, the Berkeley techno-hippie who wired Ken Kesey's bus for the Merry Prankster expeditions and who was also a synth innovator, but he'd make an instrument only for those he deemed worthy. They all pioneered capacitive-touch technology, as did Buxton, in their aural experiments.

* * *

The first device that we would recognize as a touchscreen today is believed to have been invented by Eric Arthur Johnson, an engineer at England's Royal Radar Establishment, in 1965. And it was created to improve air traffic control.

In Johnson's day, whenever a pilot called in an update to his or her flight plan, an air traffic controller had to type a five- to seven-character call sign into a teleprinter in order to enter it on an electronic data display. That extra step was time-consuming and allowed for user error.

A touch-based air traffic control system, he reckoned, would allow controllers to make changes to aircraft's flight plans more efficiently.

Johnson's initial touchscreen proposal was to run copper wires across the surface of a cathode-ray tube, basically creating a touchable TV. The system could register only one touch at a time, but the groundwork for modern touchscreens was there—and it was capacitive, the more complex kind of touchscreen that senses when a finger creates a change in capacitance, right from the start.

The touchscreen was linked to a database that contained all of the call signs of all the aircraft in a particular sector. The screen would display the call signs, "one against each touch wire." When an aircraft called to identify itself, the controller would simply touch the wire against its call sign. The system would then offer the option to input only changes to the flight plan that were allowable. It was a smart way to reduce response times in a field where every detail counted—and where a couple of incorrect characters could result in a crash.

"Of course other possible applications exist," Johnson wrote. For instance, if someone wanted to open an app on a home screen. Or had a particle accelerator to control.

For a man who made such an important contribution to technology, little is on the record about E. A. Johnson. So it's a matter of speculation as to what made him take the leap into touchscreens. We do know what Johnson cited as prior art in his patent, at least: two Otis Elevator patents, one for capacitance-based proximity sensing (the technology that keeps the doors from closing when passengers are in the way) and one for touch-responsive elevator controls. He also named patents from General Electric, IBM, the U.S. military, and American Mach and Foundry. All six were filed in the early to mid-1960s; the idea for touch control was "in the air" even if it wasn't being used to control computer systems.

Finally, he cites a 1918 patent for a "type-writing telegraph system." Invented by Frederick Ghio, a young Italian immigrant who lived in Connecticut, it's basically a typewriter that's been flattened into a tablet-size grid so each key can be wired into a touch system. It's like the analog version of your smartphone's keyboard. It would have allowed for the automatic transmission of messages based on letters, numbers, and inputs—the touch-typing telegraph was basically a pre-proto–Instant Messenger. Which means touchscreens have been tightly intertwined with telecommunications from the beginning—and they probably wouldn't have been conceived without elevators either.

E. A. Johnson's touchscreen was indeed adopted by Britain's air traffic controllers, and his system remained in use until the 1990s. But his capacitive-touch system was soon overtaken by resistive-touch systems, invented by a team under the American atomic scientist G. Samuel Hurst as a way to keep track of his research. Pressure-based resistive touch was cheaper, but it was inexact, inaccurate, and often frustrating—it would give touch tech a bad name for a couple of decades.

★ ★ ★

Back at CERN, I'm led through a crowded open hall—there's some kind of conference in progress, and there are scientists everywhere—into a stark meeting room. Stumpe takes out a massive folder, then another, and then an actual touchscreen prototype from the 1970s.

The mood suddenly grows a little tense as I begin to realize that while Stumpe is here to make the case that his technology wound up in the iPhone, Mazur is here to make sure I don't take that to be CERN's official position. They spar—politely—over details as Stumpe begins to tell me the story of how he arrived at multitouch.

Stumpe was born in Copenhagen in 1938. After high school, he joined the Danish air force, where he studied radio and radar engineering. After the service, he worked in a TV factory's development lab, tinkering with new display technologies and prototypes for future products. In 1961, he landed a job at CERN. When it came time for CERN to upgrade its first particle accelerator, the PS (Proton Synchrotron), to the *Super* PS, it needed a way to control the massive new machine. The PS had been small enough that each piece of equipment that was used to set the controls could be manipulated individually. But the PS measured a third of a mile in circumference—the SPS was slated to run 4.3 miles.

"It was economically impossible to use the old methods of direct connections from the equipment to the control room by hardwire," Stumpe says. His colleague Frank Beck had been tasked with creating a control system for the new accelerator. Beck was aware of the nascent field of touchscreen technology and thought it might work for the SPS, so he went to Stumpe and asked him if he could think of anything.

"I remembered an experiment I did in 1960 when I worked in the TV lab," Stumpe says. "When observing the time it took for the ladies to make the tiny coils needed for the TV, which was later put on the printed circuit board for the TV set, I had the idea that there might be a possibility to print these coils directly on the printed circuit board, with considerable cost savings as a result." He figured the concept could work again. "I thought if you could print a coil, you could also print a capacitor with very tiny lines, now on a transparent substrate"—like glass—"and then incorporate the capacitor to be a part of an electronic circuit, allowing it to detect a change in capacity when the glass screen was touched by a finger.... With some truth you can say that the iPhone touch technology goes back to 1960."

In March 1972, in a handwritten note, he outlined his proposal for a capacitive-touch screen with a fixed number of programmable buttons. Together, Beck and Stumpe drafted a proposal to give to the larger group at CERN. At the end of 1972, they announced the design of the new system, centered on the touchscreen and minicomputers. "By presenting successive choices that depend on previous decisions, the touch screen would make it possible for a single operator to access a large look-up table of controls using only a few buttons," Stumpe wrote. The screens would be built on cathode-ray tubes, just like TVs.

CERN accepted the proposal. The SPS hadn't been built yet, but work had to start, so its administrators set him up with what was known as a "Norwegian barrack"—a makeshift workshop erected on the open grass. The whole thing was about twenty square meters. Concept in hand, Stumpe tapped CERN's considerable resources to build a prototype. Another colleague had mastered a new technique known as ion sputtering, which allowed him to deposit a layer of copper on a clear and flexible Mylar sheet. "We worked together to create the first basic materials," he says. "That experiment resulted in the first transparent touch capacitor being embedded on a transparent surface," Stumpe says.

His sixteen-button touchscreen controls became operational in 1976, when the SPS went online. And he didn't stop working on touch tech there—eventually, he devised an updated version of his screen that would register touches much more precisely along wires arranged in an x- and y-axis, making it capable of something closer to the modern multitouch we know today. The SPS control, he says, was capable of multitouch—it could register up to sixteen simultaneous impressions—but programmers never made use of the potential. There simply wasn't a need to. Which is why his next-generation touchscreen didn't get built either.

"The present iPhones are using a touch technology which was proposed in this report here in 1977," Stumpe says, pointing to a stapled document.

He built working prototypes but couldn't gin up institutional support to fund them. "CERN told me kindly that the first screens worked fine, and why should we pay for research for the other ones? I didn't pursue the thing." However, he says, decades after, "when businesses needed to put touchscreens on mobile phones, of course people dipped into the old technology and thought, *Is this a possibility?* Industry built on the previous experience and built today what is iPhone technology."

* * *

So touch tech had been developed to manipulate music, air traffic, and particle accelerators. But the first "touch" based computers to see wide-scale use didn't even deploy proper touchscreens at all—yet they'd be crucial in promoting the concept of hands-on computing. And William Norris, the CEO of the supercomputer firm, Control Data Corporation (CDC), embraced them because he believed touching screens was the key to a digital education.

Bill Buxton calls Norris "this amazing visionary you would never expect from the seventies when you think about how computers were at the time"—i.e., terminals used for research and business. "At CDC, he saw the potential of touchscreens." Norris had experienced something of an awakening after the 1967 Detroit riots, and he vowed to use his company—and its technology—as an engine for social equality. That meant building manufacturing plants in economically depressed areas, offering day care for workers' children, providing counseling, and offering jobs to the chronically unemployed. It also meant finding ways to give more people access to computers, and finding ways to use technology to bolster education. PLATO fit the bill.

Programmed Logic for Automatic Teaching Operations was an education and training system first developed in 1960. The terminal monitors had the distinctive orange glow of the first plasma-display panels. By 1964, the PLATO IV had a "touch" screen and an elaborate, programmable interface designed to provide digital education courses. PLATO IV's screen itself didn't register touch; rather, it had light sensors mounted along each of its four sides, so the beams covered the entire surface. Thus, when you touched a certain point, you interrupted the light beams on the grid, which would tell the computer where your finger was. Norris thought the system was the future. The easy, touch-based interaction and simple, interactive navigation meant that a lesson could be beamed in to anyone with access to a terminal.

Norris "commercialized PLATO, but he deployed these things in classrooms from K through twelve throughout the state. Not every school, but he was putting computers in the classroom—more than fifteen years before the Macintosh came out—with touchscreens,"

Buxton says, comparing Norris's visionariness to Jobs's. "More than that, this guy wrote these manifestos about how computers are going to revolutionize education.... It's absolutely inconceivable! He actually puts money where his mouth is in a way that almost no major corporation has in the past."

Norris reportedly sunk nine hundred million dollars into PLATO, and it took nearly two decades before the program showed any signs of turning even a small profit. But the PLATO system had ushered in a vibrant early online community that in many ways resembled the WWW that was yet to come. It boasted message boards, multimedia, and a digital newspaper, all of which could be navigated by "touch" on a plasma display—and it promulgated the concept of a touchable computer. Norris continued to market, push, and praise the PLATO until 1984, when CDC's financial fortunes began to flag and its board urged him to step down. But with Norris behind it, PLATO spread to universities and classrooms across the country (especially in the Midwest) and even abroad. Though PLATO didn't have a true touchscreen, the idea that hands-on computing should be easy and intuitive was spread along with it.

The PLATO IV system would remain in use until 2006; the last system was shut down a month after Norris passed away.

★ ★ ★

There's an adage that technology is best when it gets out of the way, but multitouch is all about refining the way itself, improving how thoughts, impulses, and ideas are translated into computer commands. Through the 1980s and into the 1990s, touch technology continued to improve, primarily in academic, research, and industrial settings. Motorola made a touchscreen computer that didn't take off; so did HP. Experimentation with human-machine interfaces had grown more widespread, and multitouch capabilities on experimental devices like Buxton's tablet at the University of Toronto were becoming more fluid, accurate, and responsive.

But it'd take an engineer with a personal stake in the technology—

a man plagued by persistent hand injuries—to craft an approach to multitouch that would finally move it into the mainstream. Not to mention a stroke of luck or two to land it into the halls of one of the biggest technology companies in the world.

* * *

In his 1999 PhD dissertation, "Hand Tracking, Finger Identification, and Chordic Manipulation on a Multi-Touch Surface," Wayne Westerman, an electrical engineering graduate student at the University of Delaware, included a strikingly personal dedication.

> *This manuscript is dedicated to:*
> *My mother, Bessie,*
> *who taught herself to fight chronic pain in numerous and clever ways,*
> *and taught me to do the same.*

Wayne's mother suffered from chronic back pain and was forced to spend much of her day bedridden. But she was not easily discouraged. She would, for instance, gather potatoes, bring them to bed, peel them lying down, and then get back up to put them on to boil in order to prepare the family dinner. She'd hold meetings in her living room—she was the chair of the American Association of University Women—over which she would preside while lying on her back. She was diligent, and she found ways to work around her ailment. Her son would do the same. Without Wayne's—and Bessie's—tactical perseverance in the face of chronic pain, in fact, multitouch might never have made it to the iPhone.

Westerman's contribution to the iPhone has been obscured from view, due in no small part to Apple's prohibitive nondisclosure policies. Apple would not permit Westerman to be interviewed on the record. However, I spoke with his sister, Ellen Hoerle, who shared the Westerman family history with me.

Born in Kansas City, Missouri, in 1973, Wayne grew up in the small town of Wellington, which is about as close to the actual middle of

America as you can possibly get. His sister was ten years older. Their parents, Bessie and Howard, were intellectuals, a rare breed in Wellington's rural social scene. Howard, in fact, was pushed out of his first high-school teaching job for insisting on including evolution in the curriculum.

Early on, Wayne showed an interest in tinkering. "They bought him just about every Lego set that had ever been created," Hoerle says, and his parents started him on piano when he was five. Tinkering and piano, she says, are the two things that opened up his inventive spirit. They'd set up an electric train set in the living room, where it'd run in a loop, winding through furniture and around the room. "They thought, *This kid's a genius,*" Hoerle says. And Wayne was indeed excelling. "I could tell when he was five years old that he could learn faster than some of *my* peers," she recalls. "He just picked things up so much faster than everybody else. They had him reading the classics and subscribed to *Scientific American.*"

Bessie had to have back surgery, which marked the beginning of a lifelong struggle. "That's another thing that was very important about our family. A year after that, she basically became disabled with chronic pain," Hoerle says. Ellen, now a teenager, took charge of "the physical side of motherhood" for Wayne. "I had to kind of raise him. I had to keep him out of trouble."

When Ellen went off to college, it left her brother isolated. He already didn't relate particularly well to other kids, and now he had to do the household work his sister used to. "Cooking, cleaning, sorting laundry, all things he had to take over when he was eight." By his early teens, Westerman was trying to invent things of his own, working with the circuits and spare parts at his father's school. His dad bought kits designed to teach children about circuits and electricity, and Wayne would help repair the kits, which the high-school kids tore through.

He graduated valedictorian and accepted a full-ride to Purdue. There, he was struck by tendinitis in his wrists, a repetitive strain injury that would afflict him for much of his life. His hands started to hurt while he was working on papers, sitting perched in front of his computer for hours. Instead of despairing, he tried to invent his way to a solution. He took the special ergonomic keyboards made by a company called Kinesis and attached rollers that enabled him to move his hands back and forth as he typed, reducing the repetitive strain. It worked well enough that he thought it should be patented; the Kansas City patent office thought otherwise. Undeterred, Wayne trekked to Kinesis's offices in Washington, where the execs liked the concept but felt, alas, that it would be too expensive to manufacture.

He finished Purdue early and followed one of his favorite professors, Neal Gallagher, to the University of Delaware. At the time, Wayne was interested in artificial intelligence, and he set out to pursue his PhD under an accomplished professor, Dr. John Elias. But as his studies progressed, he found it difficult to narrow his focus.

Meanwhile, Westerman's repetitive strain injuries had returned with a vengeance. Some days he physically couldn't manage to type much more than a single page.

"I couldn't stand to press the buttons anymore," he'd say later. (Westerman has only given a handful of interviews, most before he joined Apple, and the quotes that follow are drawn from them.) Out of necessity, he started looking for alternatives to the keyboard. "I noticed my hands had much more endurance with zero-force input like optical buttons and capacitive touch pads."

Wayne started thinking of ways he could harness his research to create a more comfortable work surface. "We began looking for one," he said, "but there were no such tablets on the market. The touch pad manufacturers of the day told Dr. Elias that their products could not process multi-finger input.

"We ended up building the whole thing from scratch," Westerman said. They shifted the bulk of their efforts to building the new touch device, and he ended up "sidetracked" from the original dissertation topic, which had been focused on artificial intelligence. Inspiration had struck, and Wayne had some ideas for how a zero-force, multi-finger touchpad might work. "Since I played piano," he said, "using all ten fingers seemed fun and natural and inspired me to create interactions that flowed more like playing a musical instrument."

Westerman and Elias built their own key-free, gesture-recognizing touchpad. They used some of the algorithms they developed for the AI project to recognize complex finger strokes and multiple strokes at once. If they could nail this, it would be a godsend for people with RSIs, like Wayne, and perhaps a better way to input data, period.

But it struck some of their colleagues as a little odd. Who would want to tap away for an extended period on a flat pad? Especially since keyboards had already spent decades as the dominant human-to-computer input mechanism. "Our early experiments with surface typing for desktop computers were met with skepticism," Westerman said, "but the algorithms we invented helped surface typing feel crisp, airy, and reasonably accurate despite the lack of tactile feedback."

Dr. Elias, his adviser, had the skill and background necessary to translate Wayne's algorithmic whims into functioning hardware. Neal Gallagher, who'd become chair of the department, ensured that the school helped fund their early prototypes. And Westerman had received support from the National Science Foundation to boot.

Building a device that enabled what would soon come to be known as multitouch took over Westerman's research and became the topic of his dissertation. His "novel input integration technique" could recognize both single taps and multiple touches. You could switch seamlessly between typing on a keyboard and interacting with multiple fingers with whatever program you were using. Sound familiar? The keyboard's there when you need it and out of the way when you don't.

But Wayne's focus was on building an array of gestures that could replace the mouse and keyboard. Gestures like, say, pinching the pad with your finger and thumb to—okay, *cut* at the time, not *zoom*. Rotating your fingers to the right to execute an open command. Doing the same to the left to close. He built a glossary of those gestures, which he believed would help make the human-computer interface more fluid and efficient.

Westerman's chief motivator still was improving the hand-friendliness of keyboards; the pad was less repetitive and required lighter keystrokes. The ultimate proof was in the three-hundred-plus-page dissertation itself, which Wayne had multitouched to completion. "Based upon my daily use of a prototype to prepare this document," he concluded, "I have found that the [multitouch surface] system as a whole is nearly as reliable, much more efficient, and much less fatiguing than the typical mouse-keyboard combination." The paper was published in 1999. "In the past few years, the growth of the internet has accelerated the penetration of computers into our daily work and lifestyles," Westerman wrote. That boom had turned the inefficiencies of the keyboard into "crippling illnesses," he went on, arguing, as Apple's ENRI team would, that "the conventional mechanical keyboard, for all of its strengths, is physically incompatible with the rich graphical manipulation demands of modern software." Thus, "by replacing the keyboard with a multitouch-sensitive surface and recognizing hand motions...hand-computer interaction can be dramatically transfigured." How right he was.

* * *

The success of the dissertation had energized both teacher and student, and Elias and Westerman began to think they'd stumbled on the makings of a marketable product. They patented the device in 2001 and formed their company, FingerWorks, while still under the nurturing umbrella of the University of Delaware. The university itself became a shareholder in the start-up. This was years before *incubators* and *accelerators* became buzzwords—outside of Stanford and MIT, there weren't a lot of universities providing that sort of support to academic inventors.

In 2001, FingerWorks released the iGesture NumPad, which was about the size of a mousepad. You could drag your fingers over the pad, and sensors would track their movements; gesture recognition was built in. The pad earned the admiration of creative profession-als, with whom it found a small user base. It made enough of a splash that the *New York Times* covered the release of FingerWorks' second product: the $249 TouchStream Mini, a full-size keyboard replace-ment made up of two touchpads, one for each hand.

"Dr. Westerman and his co-developer, John G. Elias," the newspa-per of record wrote, "are trying to market their technology to others whose injuries might prevent them from using a computer." Thing was, they didn't have a marketing department.

Nonetheless, interest in the start-up slowly percolated. They were selling a growing number of pads through their website, and their dedicated users were more than just dedicated; they took to calling themselves Finger Fans and started an online message board by the same name. But at that point, FingerWorks had sold around fifteen hundred touchpads.

At an investment fair in Philadelphia, they caught the attention of a local entrepreneur, Jeff White, who had just sold his biotech com-pany. He approached the company's booth. "So I said, 'Show me what you have,'" White later said in an interview with Technical.ly Philly. "He put his hand on his laptop and right away, I got it… Right away I got the impact of what they were doing, how breakthrough it was." They told him they were looking for investors.

"With all due respect," White told them, "you don't have a manage-ment team. You don't have any business training. If you can find a man-agement team, I'll help you raise the rest of the money." According to White, the FingerWorks team essentially said, Well, you just sold your company—why not come run ours? He said, "Make me a cofounder and give me founder equity," and he'd work the way they did—he wouldn't take a salary. "It was the best decision I ever made," he said.

White hatched a straightforward strategy. Westerman had carpal tunnel syndrome, so his primary aim was to help people with hand disabilities. "Wayne had a very lofty and admirable goal," White said. "I just want to see it on as many systems as possible and make some money on it. So I said, 'If we sold the company in a year, you'd be OK with that?'" White set up meetings with the major tech giants of the day—IBM, Microsoft, NEC, and, of course, Apple. There was interest, but none pulled the trigger.

Meanwhile, FingerWorks continued its gradual ascent; its customer base of Finger Fans expanded and the company began collecting mainstream accolades. At the beginning of 2005, FingerWorks' iGesture pad won the Best of Innovation award at CES, the tech industry's major annual trade show.

Still, at the time, Apple execs weren't convinced that FingerWorks was worth pursuing—until the ENRI group decided to embrace multitouch. Even then, an insider at Apple at the time who was familiar with the deal tells me that the executives gave FingerWorks a lowball offer, and the engineers initially said no. Steve Hotelling, the head of the input group, had to personally call them up and make his case, and eventually they came around.

"Apple was very interested in it," White said. "It turned from a licensing deal to an acquisition deal pretty quickly. The whole process took about eight months."

As part of the deal, Wayne and John would head west to join Apple full-time. Apple would obtain their multitouch patents. Jeff White, as co-founder, would enjoy a considerable windfall. But Wayne had some reservations about selling FingerWorks to Apple, his sister suggests. Wayne very much believed in his original mission—to offer the many computer users with carpal tunnel or other repetitive strain injuries an alternative to keyboards. He still felt that Finger-Works was helping to fill a void and that in a sense he'd be abandoning his small but passionate user base.

Sure enough, when FingerWorks' website went dark in 2005, a wave of alarm went through the Finger Fans community.

One user, Barbara, sent a message to the founder himself and then posted to the group.

Just received a (very prompt) reply for my email to Wayne Westerman, in which I asked him: "Have you sold the company and will your product line be taken up and continued by another business?" Westerman wrote back: "I wish manufacturing had continued or shutdown had gone smoother, but if we all cross our fingers, maybe the basic technology will not disappear forever. :-)"

When the iPhone was announced in 2007, everything suddenly made sense. Apple filed a patent for a multitouch device with Westerman's name on it, and the gesture-controlled multitouch technology was distinctly similar to FingerWorks'. A few days later, Westerman underlined that notion when he gave a Delaware newspaper his last public interview: "The one difference that's actually quite significant is the iPhone is a display with the multi-touch, and the FingerWorks was just an opaque surface," he said. "There's definite similarities, but Apple's definitely taken it another step by having it on a display."

The discontinued TouchStream keyboards became highly sought after, especially among users with repetitive strain injuries. On a forum called Geekhack, one user, Dstamatis, reported paying $1,525 for the once-$339 keyboard: "I've used Fingerworks for about 4 years, and have never looked back." Passionate users felt that FingerWorks' pads were the only serious ergonomic alternative to keyboards, and now that they'd been taken away, more than a few Finger Fans blamed Apple. "People with chronic RSI injuries were suddenly left out in the cold, in 2005, by an uncaring Steve Jobs," Dstamatis wrote. "Apple took an important medical product off the market."

No major product has emerged to serve RSI-plagued computer users, and the iPhone and iPad offer only a fraction of the novel interactivity of the original pads. Apple took FingerWorks' gesture library and simplified it into a language that a child could understand — recall that Apple's Brian Huppi had called FingerWorks' gesture database an "exotic language" — which made it immensely popular. Yet if FingerWorks had stayed the course, could it have taught us all a new, richer language of interaction? Thousands of FingerWorks customers' lives were no doubt dramatically improved. In fact the ENRI crew at Apple might never have investigated multitouch in the first place if Tina Huang hadn't been using a FingerWorks pad to relieve her wrist pain. Then again, the multitouch tech Wayne helped put into the iPhone now reaches *billions* of people, as it's become the de facto language of Android, tablets, and trackpads the world over. (It's also worth noting that the iPhone would come to host a number of accessibility features, including those that assist the hearing and visually impaired.)

Wayne's mother passed away in 2009, from cancer. His father passed a year later. Neither owned an iPhone — his father refused to use cell phones as a matter of principle — though they were proud of their son's achievements. In fact, so is all of Wellington. Ellen Hoerle says the small town regards Wayne as a local hero.

Like his mother, Wayne had found a clever way around chronic pain. In the process, he helped, finally, usher in the touchscreen as the dominant portal to computers, and he wrote the first dictionary for the gesture-based language we all now speak.

★ ★ ★

Which brings us back to Jobs's claim that Apple invented multitouch. Is there any way to support such a claim? "They certainly did not invent either capacitive-touch or multitouch," Buxton says, but they "contributed to the state of the art. There's no question of that." And Apple undoubtedly brought both capacitive touchscreens and multitouch to the forefront of the industry.

Apple tapped a half a century's worth of touch innovation, bought out one of its chief pioneers, and put its own formidable spin on its execution. Still, one question remains: Why did it take so long for touch to become the central mode of human-machine interaction when the groundwork had been laid decades earlier? "It always takes that long," Buxton says. "In fact, multitouch went faster than the mouse."

Buxton calls this phenomenon the Long Nose of Innovation, a theory that posits, essentially, that inventions have to marinate for a couple of decades while the various ecosystems and technologies necessary to make them appealing or useful develop. The mouse didn't go mainstream until the arrival of Windows 95. Before that, most people used the keyboard to type on DOS, or, more likely, they used nothing at all.

"The iPhone made a quantum leap in terms of being the first really successful digital device that had, for all intents and purposes, an analog interface," Buxton says. He gets poetic when describing how multitouch translates intuitive movements into action: "Up until that point, you poked, you prodded, you bumped, you did all this stuff, but nothing flowed, nothing was animated, nothing was alive, nothing flew. You didn't caress, you didn't stroke, you didn't fondle. You just push. You poke, poke, poke, and it went blip, flip, flip. Things jumped; they didn't flow."

Apple made multitouch flow, but they didn't create it. And here's why that matters: Collectives, teams, multiple inventors, build on a shared history. That's how a core, universally adopted technology emerges—in this case, by way of boundary-pushing musical experimenters; smart, innovative engineers with eyes for efficiency; idealistic, education-obsessed CEOs; and resourceful scientists intent on creating a way to transcend their own injuries.

"The thing that concerns me about the Steve Jobs and Edison complex—and there are a lot of people in between and those two are just two of the masters—what worries me is that young people who are being trained as innovators or designers are being sold the Edison myth, the genius designer, the great innovator, the Steve Jobs, the Bill Gates, or whatever," Buxton says. "They're never being taught the notion of the collective, the team, the history."

* * *

Back at CERN, Bent Stumpe made an impressively detailed case that his inventions had paved the way for the iPhone. The touchscreen report was published in 1973, and a year later, a Danish firm began manufacturing touchscreens based on the schematic. An American magazine ran a feature about it, and hundreds of requests for information poured in from the biggest tech companies of the day. "I went to England, I went to Japan, I went all over and installed things related to the CERN development," Stumpe says. It seems entirely plausible that Stumpe's touchscreen innovations were absorbed into the touchscreen bloodstream without anyone giving him credit or recompense. Then again, as with most sapling technologies, it's almost impossible to tell which was first, or concurrent, or foundational.

After the tour, Stumpe invites me back to his home. As we leave, we watch a young man slinking down the sidewalk, head bent over his phone. Stumpe laughs and shakes his head with a sigh as if to say, *All this for that?*

All this for that, maybe. One of the messy things about dedicating your life to innovation—real innovation, not necessarily the buzzword deployed by marketing departments—is that, more often than not, it's hard to see how, or if, those innovations play out. It may feed into a web so thick any individual threads are inscrutable, and it may contribute to the richness of the ideas "in the air." Johnson, Theremin, Norris, Moog, Stumpe, Buxton, Westerman—and the teams behind them—who's to say how and if the iPhone's interface would feel without any of their contributions? Of course, it takes another set of skills entirely to develop a technology into a product that's universally desirable, and to market, manufacture, and distribute that product—all of which Apple happens to excel at.

But imagine watching the rise of the smartphone and the tablet, watching the world take up capacitive touchscreens, watching a billionaire CEO step out onto a stage and say his company invented them—thirty years after you were certain you proved the concept. Imagine watching that from the balcony of your third-floor one-bedroom apartment in the suburbs of Geneva that you rent with your pension and having proof that your DNA is in the device but finding that nobody seems to care. That kind of experience, I'm afraid, is the lot of the majority of inventors, innovators, and engineers whose collective work wound up in products like the iPhone.

We aren't great at conceiving of technologies, products, even works of art as the intensely multifaceted, sometimes generationally collaborative, efforts that they tend to be. Our brains don't tidily compute such ecosystemic narratives. We want eureka moments and justified millionaires, not touched pioneers and intangible endings.

ii: Prototyping

First draft of the one device

Jony Ive had finally decided that the time was right. Maybe Jobs was in an unusually good mood when he dropped by the ID studio for one of his frequent visits. Maybe Ive felt the demos that the engineers and UI wizards had cooked up—the pinch to zoom, the rotating maps—were as good as they were going to get on a wonky, stitched-together rig. Either way, one day in the summer of 2003, Ive led Jobs into the user testing facility adjacent to his design studio, where he unveiled the ENRI project and gave him a hands-on demonstration of the powers of multitouch.

"He was completely unimpressed," Ive said. "He didn't see that there was any value to the idea. And I felt really stupid because I had perceived it to be a very big thing. I said, 'Well, for example, imagine the back of a digital camera. Why would it have a small screen and all of these buttons? Why couldn't it be all display?' That was the first application I could think of on the spot, which is a great example of how early on this was."

"Still he was very, very dismissive," Ive said.

In Jobs's defense, it was a table-sized contraption with a projector pointed at a white piece of paper. The Apple CEO looked for products, not science projects.

According to Ive, Jobs spent the next few days thinking it over, and evidently changed his mind. Soon, in fact, he decided that he loved it. Later—as we saw last chapter—he would publicly announce that Apple invented it. And then would go on to tell the journalist Walt Mossberg that he'd come up with the idea of doing a multitouch tablet himself. "I'll actually tell you kind of a secret," Jobs said. "I actually started on the tablet first. I had this idea of being able to get rid of the keyboard, to type on a multitouch glass display." Yet Jobs likely didn't even know about the touch-tablet project until Jony had given him a demonstration. And he'd initially rejected it.

"When he saw the first prototype," Strickon says, "I think the quote was either 'This thing is only good for reading my email on the toilet' or the other thing I heard was that he wanted a device that he could [use to] read email on the toilet. It came out both ways." Regardless, that became the product spec: Steve wanted a piece of glass he could read his email on.

At one point, the ENRI group was standing around the ID studio, when Greg Christie blew in. He'd been meeting with Jobs on a regular basis about multitouch. "Now what's the latest from Steve on this?" someone asked.

"Well," he said, "First thing everyone needs to note is: Steve invented multitouch. So everybody go back and change your notebooks." And then he grinned.

They rolled their eyes and laughed—that was pure Steve Jobs. Even now, Huppi's amused when the Mossberg incident comes up. "Steve said, 'Yeah, I went to my engineers and said "I want a thing that does this this and this"'—and that's all total bullshit because he had never asked for that." No one heard Steve talk about multitouch before he saw the ENRI team's demo. "As far as I know, Jony showed him the demo of multitouch and then it was clicking in his mind.... Steve does this, you know: He comes back later and it's his idea. And no one's going to convince him otherwise." Huppi laughs. He doesn't seem bitter; it was a fact of life at Jobs-led Apple. "And that's fine."

★ ★ ★

Jobs's approval raised the profile of the project, and, unsurprisingly, stirred up interest inside the company. "These meetings were a lot bigger now," Huppi says. A project was greenlit to translate the fragments, ideas, and ambitions of the ENRI experiments into a product. The hardware effort to transform the rig into a working prototype—which at the time was a multitouch tablet—was given a code name: Q79. The project went on lockdown.

And there was still a long way to go.

The rig still relied on a plastic FingerWorks pad, for one. "The next question was, How the hell would you do it on a clear screen?" Huppi says. "And we had no idea how that was going to happen."

That's because the FingerWorks pad was absolutely loaded with chips. "For every little five-millimeter-by-five-millimeter patch, there was an electrode going to a chip—and so there were a lot of chips to cover a whole device of that size," Huppi says. That was fine for an opaque black pad where you could hide them—but how could they ever do that on glass with a screen underneath? Huppi wondered.

So, Josh Strickon hit the books, reading up on the touch tech literature, digging through papers and published experiments and tinkering with alternatives. Research at Apple wasn't always easy; Steve Jobs had shut down the company library, which used to provide engineers and designers with an archival resource, after his return. Still, Strickon was resolute: "There had to be a better way."

Smarter Skin

Soon, he had some good news. Strickon thought he'd found a solution that might let them do multitouch on glass without having to deal with an avalanche of chips.

"I found Sony SmartSkin," Strickon says. Sony was in the process of becoming one of Apple's chief competitors—it was losing market share in portable music players to Apple's iPods. Sony had been digging into capacitive sensing too. "This paper from Sony implied that you could do true multitouch with rows and columns," Strickon says. It meant a lattice of electrodes laid out on the screen could do the sensing.

Josh Strickon considers this to be one of the most crucial moments in the course of the project. The paper, he says, presented a "much more elegant way" to do multitouch. It just hadn't been done on a transparent surface yet. So, tracing the outline of Sony's SmartSkin, he patched together a DIY multitouch screen. "I built that first pixel with a sheet of glass and some copper tape. That is what kicked the whole thing off," he says, bothered not one iota that the method was borrowed from the competition. "I came from a research background where you look at what is in the field." That approach, and the prospect of building new products based on another company's research, spooked Apple's legal team. "Once things started getting going with multitouch, the lawyers instructed us not to do those sorts of searches anymore," Strickon says, annoyed at the memory. "I am not sure how you are expected to innovate without having an understanding about what was done before."

Regardless, that pixel was tangible proof that you wouldn't need to load up a device with chips to do multitouch. The input team now had to expand that lonely pixel into a full, tablet-size panel. So they went on a shopping spree.

"We scrambled—we would grab some parts from RadioShack or wherever we had to get them, and we cooked this thing up on glass," Huppi says. "It was a piece of glass with a couple of copper electrodes taped in there, and I mean totally cooked up, breadboard-style." Breadboards are what engineers use to prototype electronics; they started out as actual breadboards, of the wooden, yeast-handling variety, on which radio tinkerers soldered wires and evolved into a standard tool engineers use to experiment.

Touch had never been done quite like this before. The team built three of those breadboards—large, poker-table-size arrays with their guts splayed out on top—to prove the rig would be capable of registering real interactions. There's one known prototype left, evidence of the earliest stages of the iPhone's evolution, and it's tucked away in an office at Apple. I was shown a rare picture of that first breadboard— it looked like raw chipboards often do, like a green sound-mixing board with an inlaid screen, surrounded by a rigid sea of circuits.

In order to understand exactly how a user's hand was interacting with the touchscreen, Strickon programmed a tool that created a visualization of the palm and fingers hitting the sensors in real time. "It was kind of like those bed of nails that you put your hand into," Huppi says, that "creates this three-D image of your hand on the other side." They called it the Multitouch Visualizer. "That was literally like the first thing we put on the screen," Strickon says. According to Huppi, it's still used to monitor touch sensors at Apple today.

Strickon took the opportunity to mess around with the musical capabilities of the new device too, and wrote a program that transformed the touchpad into a working theremin. Moving his hand left or right modulated pitch, while moving it up or down controlled volume. The ancestor of the iPhone could be played like a Russian proto-synth before it could do much of anything else. "We had some fun doing goofy stuff like that," Huppi says.

The prospect was tantalizing: Not only could multitouch power a new kind of tablet that users could directly manipulate — one that was fun, efficient, and intuitive — but it could work on a whole suite of trackpads and input mechanisms for typical computers too. A phone was still the furthest thing from their minds.

* * *

The hardware team was putting together a working touchscreen. Jobs was enthusiastic. Industrial Design was looking into form-factor ideas. And the tablet was going to need a chip to run the touch sensor software that Strickon had cooked up.

The team had never designed a custom chip before, but their boss Steve Hotelling had, and he pushed forward. "He just said, 'Yeah, no problem, we'll just get some bids out there we'll get a chip made.... It's going to take about a million bucks; in eight months we'll have a chip,'" Huppi says.

They settled on the Southern California chipmaker Broadcom. In an unusual move, Apple invited the company's reps to come in and see the "magic" themselves. "I don't think it's ever been done since,"

Huppi says. Hotelling thought that if the outside contractors saw the demo in action, they'd be excited by it and do the work better and faster. So, a small Broadcom team was led into the testing lab. "It was just this amazing moment to see how excited these guys were," Huppi says. "In fact, one of those guys that's now at Apple, he came from that team, and he still remembers that day."

But it would be months before the chip was done.

"Meanwhile, there was again a big driving force to build more 'Looks like, feels like' prototypes," Huppi says. As the project grew, more and more people wanted their hands on it. The team knew that if they didn't deliver, executives might lose interest. "We ended up building this tethered display, which looked like an iPad but plugged into your computer," Strickon says. "That's one of the things I pushed for—like, we've got to build stuff to put in their hands."

The first round of proto-tablets they delivered weren't exactly stellar, however.

"There were prototypes that were built that were basically like tablets, Mac tablets—things that could barely have more than an hour's battery life." Strickon says with a laugh. "It had no usefulness." Which was a bit more generous than Bas Ording's assessment. He, after all, had to use them to work on the UI. "The thing would overheat in that enclosure, and the battery life was like two minutes," he says with a laugh.

The team built fifty or so prototypes—thick, white, tablet-looking ancestors of the iPad. That way, software designers could just jack the touchpad into Mac software without sacrificing any performance or power, and the UI team could keep working on perfecting the interface.

Culture Clash

As the project drew on, some members of the team chafed under Apple's rigid culture. Strickon, for one, wasn't used to the corporate hierarchy and staid atmosphere—he was an ambitious, unorthodox researcher and experimenter, recall, and still fresh out of MIT. His boss, Steve Hotelling, chastised him for interrupting superiors (like Hotelling) at meetings; Strickon, meanwhile, shot back that Hotelling was a square "company man." Worse was the old boys' club that had a stranglehold on decision making, he says. "A lot of people had been there for a long, long time. People like [senior vice president of marketing] Phil Schiller, they've been around forever—that club was already kind of established." He felt like his ideas were dismissed at meetings. "It was like, 'I see how it is out here—not everybody's supposed to have ideas.'"

Outside of the project, Strickon was lonely and despondent. "I was trying to meet people, but it really wasn't working. . . . Even HR was trying to help me out," he says, "introducing me to people."

Ording's good nature was tested from time to time too. He'd been sitting in weekly meetings with the CEO but often found Jobs's mean streak too much to handle. "There was a period of time," Ording says, "that for a couple months or so, half a year or whatever, I didn't go to the Steve meetings." Jobs would chew out his colleagues in a mean-spirited way that made Ording not want to participate. "I just didn't want to go. I was like, 'No, Steve's an asshole.' Too many times he would be nasty for no good reason," he says. "No one understood, because most people would die to go to these meetings—like, 'Oh, it's Steve.' But I was tired of it."

Amid all that, the team had no clear idea of what, exactly, they were trying to make. A fully powered Mac you could touch? A mobile device on a completely different operating system? Try to transport your brain to the distant reaches of the turn of the twentieth century—there were no touchscreen tablets in wide use; they were still more familiar as *Star Trek*-esque fictions than actual products.

The ENRI team was in uncharted territory.

UI for U and I

Freed from the confines of point-and-click, Ording and Chaudhri continued to embrace the possibilities of direct manipulation. Their demos became more finely tuned, more ambitious. "You had the feeling that you could come up with whatever—it's great for UI design. You have almost a clean slate," Ording says.

"For the first time, we had something that's like direct manipulation, as opposed to what we *used* to call direct manipulation—which was clicking on an icon, but there was a mouse," Greg Christie says. There was still one extra intermediary between you and the computer. "It's like operating a robot." Now, it wasn't: It was hand-to-pixel contact.

That sort of direct manipulation meant the rules that governed point-and-click computing were out the window. "Because we could start from scratch, and we could do whatever, so there was much more animation and nice transitions to get the whole thing a certain feel that people hadn't seen before," Ording says. "Combined with actual multitouch, it was even more magical. It was all very natural in some ways, its own little virtual reality."

To craft that new virtual reality, Ording followed his instincts toward playful design. A longtime player and admirer of video games, he baked in gamelike tendencies to try to make even the most insignificant-seeming interactions feel compelling. "My interest is in how you can make something that's fun to use but also of course functional," Ording says, "like the scrolling on the iPhone with the little bouncing thing on the end. I describe it as playful, but at the same time it's very functional, and when it happens, it's like, 'Oh, this is kind of fun,' and you just want to do it again, just to see the effect again."

Early video games like Pac-Man or Donkey Kong—the kind a young Bas Ording played growing up—were highly repetitive affairs that hooked players with a series of tiny carrots and rewards. With an extremely limited repertoire of moves—go up, go down, jump, run—getting to the next level was all about mastering the narrow controls. When you did, it became satisfying to move fluidly through a level—and then to rack up your score and discover what lay in store with the next one.

"Games are all about that, right? They make you want to keep playing the game," Ording says. So his design sensibility makes you want to discover the next thing, to tinker, to explore. "For some reason, software has to be boring. I never got that, why people wouldn't put the same kind of attention to the way things move or how you interact with it to make it a fun experience, you know?" They'd come up with the blueprint for a new brand of computing, laying the basic foundation with pleasant, even addictive, flourishes. Ording's design animations, embedded since the earliest days, sharpened by Chaudhri's sense of style, might be one reason we're all so hooked on our smartphones.

And they did it all on basic Adobe software.

"We built the entire UI using Photoshop and Director," Chaudhri says, laughing. "It was like building a Frank Gehry piece out of aluminum foil. It was the biggest hack of all time." Years later, they told Adobe — "They were fucking floored."

Glitches

By the end of 2003, Apple still hadn't completed its rebound into a cash-rich megacompany. And some employees were beset by standard-issue workplace woes: low pay and bad office equipment. Inventing the future is less fun with stagnant wages. "Money was pretty tight at that time," Strickon says. Salaries were "pretty low; people weren't so happy and not getting raises and not getting bonuses." Strickon's and Huppi's computers were buggy, and frequently malfunctioning. They couldn't manage to get Apple to replace their Macs.

And it turned out they'd need more than just Macs for touch prototyping. "It was funny because we had to buy a PC," Strickon says. "All the firmware tools were for Windows. So we ended up building a PC out of parts...but that was easier to get than a working Mac."

As Q79 built momentum, the marketing department remained skeptical about the product, even with Jobs on board. They couldn't quite imagine why anyone would want to use a portable touch-based device.

Strickon recalls one meeting where tempers flared after the younger engineers tried to make the case for the tablet and saw their ideas shot down. They were gathered in the ID studio with Tim Bucher, one of the first Apple executives to throw his weight behind the project. One of the reps from marketing got so incensed, Bucher had to stop the meeting. He said, "'Look, anyone here is allowed to have an idea,'" Strickon recalls. "That was the biggest problem.... We were trying to define a new class of computing device, and no one would really talk to us about it," Strickon says, to understand what they were trying to do.

The marketing department's ideas for how to sell the new touch-based device didn't exactly inspire confidence either. They put together a presentation to show how they could position the tablet to sell to real estate agents, who could use it to show images of homes to their clients. "I was like, 'Oh my God, this is so off the mark,'" Strickon says.

Jobs had increased secrecy for the Q79 project—whenever products were moved around the company, they had to be covered in black cloth—and that was becoming burdensome too. "How can you communicate on projects like this if you can't trust your employees?" Strickon asks.

Few events demonstrate the paradox of working on a secret Jobs-backed project at Apple quite as well as the innovation award the Q79 team was given for—well, it couldn't say exactly.

From time to time, Apple's entire hardware division would gather for an all-hands meeting. "Every meeting, they would give out an award," Strickon says, for quality, performance, and so on. At one meeting, deep into the touchscreen project's development, Tim Bucher, the VP of Mac Engineering, stood up and delivered a speech. "With a total straight face, he says, 'We're giving out a new award—for innovation,'" Strickon recalls. He brought the Q79 team up on stage and gave them trophies: life-size red polished apples made of stone. He wouldn't, or couldn't, say anything else about it. "They literally said nothing. Nothing," Strickon says. "They're giving this team an award and couldn't tell you what it was."

Imagine the polite applause and raised eyebrows in Cupertino as a number soup codename was rewarded for innovating something that no one else was allowed to hear about. It'd be like the Academy giving an Oscar to a new Coen brothers film that only its members had been allowed to see.

"Classic internal secrecy bullshit," Huppi says. "I still have that award somewhere."

★ ★ ★

Meanwhile, the input team searched for a supplier that could churn out the panel tech at quality and scale. There were late-night conference calls and trips to Taiwan. The market for LCD screens was going crazy at that point, Strickon says, so finding time on a production line was tough.

When they finally did, there was a major issue. "We got back our first panels from WinTech to test," Strickon says, "and you stuck it on the screen, and the next thing you know, you had a plaid screen."

The touch sensors were creating an obtrusive highway of electrodes over the tablet surface. So Strickon hid the traffic with another invention: he whipped up a "dummy pattern in between that would make it appear like it was a uniform, solid sheet." That was one of the key touch patents to be developed during the process, though at the time Apple's legal team rejected it, Strickon recalls. "Once [the iPhone] started taking off, they went, 'Oh, we need to revisit this!'"

Untouched

A chip was cooking. Multitouch technology was working on glass. Dozens of tablet prototypes were circulating around the Infinite Loop. But just as the tablet program should have been hitting critical mass, it was ensnared by a series of setbacks.

First, it was unclear what the software was going to look like — what operating system the touch device was going to run on and so forth. "I guess we got a little stuck with where the project was going to go," Ording says. "There was no iOS at that point. Just a bunch of weird prototype demos that we built."

"There was no product there," Christie says. "Bas had a couple of demos, one was twisting this image with two fingers and other was scrolling a list. That was all lacking a compelling virtue. It was like, okay—why? There was always a little skepticism....Apple's trackpad was so good at that point compared to the competition."

Second, it was fast becoming obvious that the tablet would be expensive.

"I remember one particular meeting where we were all standing around one of the ID tables and we decided to ask everybody, 'What would you use this thing for and how much would you be willing to pay for it?'" Huppi recalls. "Most of us were like, 'Well, I guess we'd use it to, like, look at pictures, and maybe surf the web if I'm sitting on the couch, maybe. But I don't really have a reason for email because it really wouldn't have a good keyboard.'" There seemed to be a creeping uncertainty. "The bottom line was, everybody would be willing to pay maybe five to six hundred dollars for it."

The problem was that the materials were putting the device in the thousand-dollar range, basically the same cost as a laptop. "And I think that's when Steve made that call; Steve Jobs was like, 'We can't sell this—it's too expensive,'" Huppi says.

Finally, and not least, Jobs had fallen seriously ill, and he would take multiple months off in 2004 to have long-overdue surgery to remove a malignant tumor on his pancreas. "Steve getting sick the first time, that sort of stopped things in the tracks," Strickon says. "Nothing was happening when Steve was out. It was just completely odd."

And so Q79 began to sputter.

Strickon grew frustrated with the project that seemed to be going nowhere. "There were so many hurdles to try to get people on board," he says. He watched the marketing department waffle. He listened to fruitless debate in Jobs's absence.

And he reached a breaking point. Upset by the lack of progress, the uncertainty of the project's future, and the impedance of management, Strickon was burned out. At the end of the day, he just wanted to build things.

Huppi says, "He told me something like, 'These guys don't really want to do this,' and he was just kind of getting ticked off and didn't think Apple was serious about it. So he kind of bugged out." He quit.

Josh Strickon left Apple believing the touch project would never come to fruition. He doesn't regret anything about leaving, except maybe selling his stock. "It was fun stuff, but it was also like, well, I was always interested in getting stuff out there. Not doing something in a corner that nobody sees."

The iPhone

The project languished until the end of 2004 when an executive decision came down. Jobs had decided Apple needed to do a phone.

"I got a call from Steve," Ording says. "'We're gonna do a phone. There's gonna be no buttons. Just a touchscreen.' Which was big news."

But it was bittersweet for the hardware team, who had hoped to turn their multitouch tech into a suite of input devices that used the same cybernetic language. "It was classic Steve Jobs," Huppi says. "'Drop everything else. We're doing the phone.' . . . Forget about all that other stuff. A lot of us were kind of bummed out because we were like, 'A phone? Like, really?'"

At first, it seemed like their work was getting downsized. "But this is where, again, Steve Jobs had to give us that vision. And he was like, 'No, it's perfect for the phone.'" For one thing, its small size would reduce accidental touches. For another, it would help move the touch tech into the marketplace. "It's brilliant in the phone market," Huppi says. "It's sort of subsidized by the carriers. You can have this thing that's eight hundred bucks selling for two hundred because they know they're going to have you hooked on it."

Jobs would soon pit the iPod team against a Mac software team to refine and produce a product that was more specifically phone-like. The herculean task of squeezing Apple's acclaimed operating system into a handheld phone would take another two years to complete.

Executives would clash; some would quit. Programmers would spend years of their lives coding around the clock to get the iPhone ready to launch, scrambling their social lives, their marriages, and sometimes their health in the process.

But it all had been set into motion years before. The concept of the iPhone wasn't the product of Steve Jobs's imagination—though he would fiercely oversee, refine, and curate its features and designs—but of an open-ended conversation, curiosity, and collaboration. It was a product born of technologies nurtured by other companies and then ingeniously refined by some of Apple's brightest minds—people who were then kept out of its public history.

Huppi likens it to Jobs's famous visit to Xerox PARC, when they first saw the GUI, the windows and menus that would dominate computer user interfaces for the coming decades. "It was like that... this strange little detour that turned into this big thing that's been highly influential, and it's kind of amazing that it worked out," Huppi says. "Could have just as well not, but it did."

Thanks to the ENRI group's strange little detour, the prototype of the UI you use more than any other—through your smartphone's home screen, a grid of icons that open with a touch, to be swiped, pinched, or tapped—had been brought to life.

"It's like water now," Imran Chaudhri says, "but it wasn't always so obvious."

★ ★ ★

In fact, it's still not entirely obvious. The iPhone UI may be ubiquitous, but running that water only *looks* easy. A vastly complex system sits behind the iPhone's multitouchable, scratchproof screen. This next section explores the hardware—the tiny battery, camera, processor, Wi-Fi chip, sensors, and more—that powers the one device.

CHAPTER 5

Lion Batteries

Plugging into the fuel source
of modern life

Chile's Atacama Desert is the most arid place on Earth apart from the freeze-dried poles. It doesn't take long to feel it. The parched sensation starts in the back of your throat, then moves to the roof of your mouth, and soon your sinuses feel like an animal skin that's been left under the desert sun for a week. Claudio, at the wheel, is driving me and my fixer Jason south from Calama, one of Chile's largest mining towns; the brown-red crags of the Andes loom outside our pickup's windows.

We're headed to Salar de Atacama, home to the largest lithium mine in the world. SQM, or Sociedad Química y Minera de Chile, or the Chemical and Mining Society of Chile, is the formerly state-owned, now-son-in-law-of-a-former-dictator-owned, mining company that runs the place. It's the leading producer of potassium nitrate, iodine, and lithium, and officials have agreed to let me and Jason take a private tour.

Atacama doesn't *look* ultradry; in the winter, snowcapped mountains are visible in the distance. But the entire forty-one-thousand-square-mile high desert receives an average of fifteen millimeters (about half an inch) of rain a year. In some places, it's less. There are weather stations here that have not registered rainfall in over a century of record-keeping.

Hardly anything lives in the most water-scarce regions of the Atacama, not even microbes. We stop at one of the most famously barren zones: the Valley of the Moon. It resembles Mars to such a degree that NASA used the region to test its Red Planet–bound rovers, specifically the equipment they use to search for life. And we have this barren, unearthly place to thank for keeping our iPhones running.

Chilean miners work this alien environment every day, harvesting lithium from vast evaporating pools of marine brine. That brine is a naturally occurring saltwater solution that's found here in huge underground reserves. Over the millennia, runoff from the nearby Andes mountains has carried mineral deposits down to the salt flats, resulting in brines with unusually high lithium concentrations. Lithium is the lightest metal and least dense solid element, and while it's widely distributed around the world, it never occurs naturally in pure elemental form; it's too reactive. It has to be separated and refined from compounds, so it's usually expensive to get. But here, the high concentration of lithium in the salar brines combined with the ultradry climate allows miners to harness good old evaporation to obtain the increasingly precious metal.

And Atacama is absolutely loaded with lithium — Chile currently produces a full third of the world's supply and holds a quarter of its total proven reserves. Thanks to Atacama, Chile is frequently called the "Saudi Arabia of lithium." (Then again, many, many nations could be called the "Saudi Arabia of lithium" — neighboring Bolivia has even more, but it's not mining it — yet.)

Lithium-ion batteries are the power source of choice for laptops, tablets, electric cars, and, of course, smartphones. Lithium is increasingly described as "white petroleum" by those who recognize its key place in industry. Between 2015 and 2016, lithium doubled in value because projected demand shot through the roof.

Although other mines are being developed, the best place on Earth to get lithium is nestled right here in the Chilean highlands.

As we drive, I spot a cross surrounded by flowers, photographs, and little relics on the side of the road. Then another, and another.

"Yes, this is known as Ruta del Muerte," Claudio, our driver, tells us. "Families, they don't know the roads. They get tired and drive off. Or truckers who drive too long."

The way to the stuff that makes our iPhone batteries possible is down the road of death.

* * *

Lithium-ion batteries were first pioneered in the 1970s because experts feared humanity was heading down a different, more literal, road of death due to its dependence on oil. Scientists, the public, and even oil companies were desperate for alternatives. Until then, though, batteries had been something of a stagnant technology for nearly a hundred years.

The first true battery was invented by the Italian scientist Alessandro Volta in 1799 in an effort to prove that his colleague Luigi Galvani had been wrong about frog power. Galvani had run currents of electricity through dead frogs' nervous systems—the series of experiments that would inspire Mary Shelley's *Frankenstein*—and had come to believe the amphibians had an internal store of "animal electricity." He'd noticed that when he dissected a leg that was hung on a brass hook with an iron scalpel, it tended to twitch. Volta thought that his friend's experiments were actually demonstrating the presence of an electrical charge running through the two different metal instruments via a moist intermediary. (They'd both turn out to be right—living muscle and nerve cells do indeed course with bioelectricity, *and* the fleshy frog was serving as an intermediary between electrodes.)

A battery is basically just three parts: two electrodes (an anode with a negative charge and a cathode with a positive charge) and an

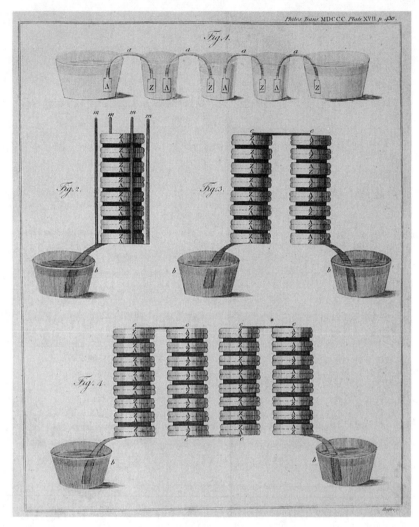

An early voltaic pile, 1793

electrolyte running between them. To test his theory, Volta built a stack of alternating zinc and copper pieces with brine-soaked cloth sandwiched between each of them. That clumsy pile was the first battery.

And it worked like most of our modern batteries do today, through oxidation and reduction. The chemical reactions cause a buildup of electrons in the anode (in Volta's pile, it's the zinc), which then want to jump to the cathode (the copper). The electrolyte—whether it's brine-soaked cloth or a dead frog—won't let it. But if you connect the battery's anode and cathode with a wire, you complete the circuit, so the anode will oxidize (lose electrons), and those electrons will travel to the cathode, generating electrical current in the process.

Expanding on Volta's concept, John Frederic Daniell created a battery that could be used as a practical source of electricity. The Daniell cell rose to prominence in 1836 and led to, among other things, the rise of the electric telegraph.

Since then, battery innovation has been slow, moving from Volta's copper-and-zinc electrodes to the lead-acid batteries used in cars to the lithium-based batteries used today. "The battery's very simplicity—its remarkably small number of parts—has both helped and hindered the efforts of scientists to improve on Volta's creation," Steve LeVine writes in *The Powerhouse.* "In 1859, a French physicist named Gaston Planté invented the rechargeable lead-acid battery," which used lead electrodes and an electrolyte of sulfuric acid. "Planté's structure went back to the very beginning—it was Volta's pile, merely turned on its side.... The Energizer, commercialized in 1980," he notes, "was a remarkably close descendant of Planté's invention. In more than a century, the science hadn't changed." Which is a little shocking, because the battery remains one of the largest silent forces that shape our experiences with technology.

But the oil shocks of the 1970s—where oil embargoes sent prices skyrocketing and crippled economies—along with the advent of a new hydrogen battery for what Ford billed as the car of the future, gave the pursuit of a better battery a shot in the arm.

* * *

Many consider it a travesty that the inventors of the lithium-ion battery haven't yet won a Nobel Prize. Not only does the li-ion battery power our gadgets, but it's the bedrock of electric vehicles. It's somewhat ironic, then, that it was invented by a scientist employed by the world's most notorious oil company.

When Stan Whittingham, a chemist, did his postdoc at Stanford in the early 1970s, he discovered a way to store lithium ions in sheets of titanium sulfide, work that resulted in a rechargeable battery. He soon received an offer to do private research into alternative energy technologies at Exxon. (Yes, Exxon, a company famous today for its efforts to cast doubt on climate change and for vying with Apple for the distinction of world's largest corporation.)

Environmentalism had swept into public consciousness after the publication of Rachel Carson's *Silent Spring* (which exposed the dangers of DDT), the Santa Barbara oil spill, and the Cuyahoga river fire. Ford moved to address complaints that its cars were polluting cities and sucking down oil by experimenting with cleaner electric cars, which instilled spark and focus to battery development. Meanwhile, it appeared that oil production had begun to peak. Oil companies were nervously eyeing the future and looking for ways to diversify.

"I joined Exxon in 1972," Whittingham tells me. "They had decided to be an energy company, not just a petroleum and chemical company. They got into batteries, fuel cells, solar cells," he says, and "at one point they were the largest producer of photovoltaic cells in the United States." They even built a hybrid diesel vehicle, decades before the rise of the Prius.

Whittingham was given a near-limitless supply of resources. The goal was "to be prepared, because the oil was going to run out."

His team knew that Panasonic had come up with a nonrechargeable lithium battery that was able to power floating LED lights for night fishermen. But those batteries could be cooled off by the ocean, an important benefit, since lithium is highly volatile and prone to generating lots of heat in a reaction.

If a battery was going to be useful to anyone who didn't have a massive source of free coolant at the workplace, it couldn't run too hot. Lithium or no, batteries can overheat if too many electrons come spilling out of the anode at once, and at the time, there was only one way out for those electrons—through the circuit. Whittingham's team changed that.

"We came up with the concept of intercalation and built the first room-temperature lithium rechargeable cells at Exxon," Whittingham says. Intercalation is the process of inserting ions between layers in compounds; lithium ions in the anode travel to the cathode, creating electricity, and since the reaction is reversible, the lithium ions can travel back to the anode, recharging the battery.

That's right—the company that spent much of 2015 and 2016 making headlines for its past efforts to silence its own scientists' warnings about the real and pressing threat of climate change is responsible for the birth of the battery that's used in the modern electric car.

"They wanted to be the Bell Labs of the energy business," Whittingham says. Bell Labs was still widely celebrated for developing the transistor, along with a spate of other wildly influential inventions. "They said, 'We need electric vehicles—let's put ourselves out of business and not let someone else put us out of business.'"

"For six decades, non-rechargeable zinc-carbon had been the standard battery chemistry for consumer electronics," LeVine writes. "Nickel-cadmium was also in use. Whittingham's brainchild was a leap ahead of both. Powerful and lightweight, it could power much smaller portable consumer electronics (think the iPod versus the Walkman)—if it worked."

The battery breakthrough sent a jolt of excitement through the division. "I was called into New York to explain to a committee on the Exxon board what we were doing and how impactful it might be," Whittingham tells me. "They were very interested."

There was a problem, however: his battery kept catching fire. "There were some flammability issues," Whittingham says. "We had several fires, mostly when we pulled them apart." Plus, it was difficult and expensive to manufacture, and it literally stank.

Thanks to the flames, the smell, and the receding of the oil crisis, Exxon never became a pioneer in electric vehicles, battery technology, or alternative energy. It doubled down on oil instead. But Whittingham's work was continued by the man who would make the consumer-electronics boom possible.

* * *

Unlike the region that envelops it, the Salar de Atacama isn't exactly beautiful. *But it's certainly striking,* I thought to myself, squinting at the salmon-colored mountains on a flat sea of thorny, twisting, dust-swept salt crystals. It looks like a dirt-swept, dry coral reef.

Those crystals would be pure white if the wind didn't blow dirt down from the mountains, says Enrique Peña, the chief engineer of the lithium mining operation at Atacama. And the fields stretch on as far as you can see.

"I imagine a Spanish conquistador was riding his horse through Chile, got here, and said, 'What the hell is this?'" Peña says. It's fifty square kilometers of nothing but arid brine. Peña is an affable young man in his mid-thirties with a shock of beard and a means-business look that easily breaks into a friendly smile. He rose quickly through the ranks at SQM, where he watches over what he affectionately refers to as "my ponds." Every week, he commutes from Santiago, where his family lives, to a lonely outpost in the high desert.

The mining operation itself, smack-dab in the middle of the salt desert, is unusual. There's no entrance carved out of rock, no deepening pit into the earth. Instead, there's a series of increasingly electric-colored, massive brine-filled evaporating pools that perfectly reflect the mountains that line the horizon. They're separated by endless mounds of salt — the by-product of the mining effort.

Underneath all that encrusted salt, sometimes just one to three meters below, there's a giant reservoir of brine, a salty solution that boasts a high concentration of lithium.

The SQM reps escort us to a lavish base camp where mining executives stay while they're visiting the site. Imagine a tiny five-star hotel with ten or so rooms and a private chef plunked down in the weird alien desert. Ground zero for the modern battery.

The perfect place, I think, to phone its inventor.

* * *

When I tell John Goodenough that I'm calling him from a lithium mine in the Atacama Desert, he lets loose a howling hoot. Goodenough is a giant in his field—he spearheaded the most important battery innovations since Whittingham's lithium breakthrough—and that laugh has become notorious. At age ninety-four, he still heads into his office nearly every day, and he tells me he's on the brink of one last leap forward in the rechargeable world.

Goodenough, an army vet who studied physics under Edward Teller and Enrico Fermi at the University of Chicago, began his career at MIT's Lincoln Laboratory, investigating magnetic storage. By the mid-1970s, like Whittingham, he was moved by the energy crisis to research energy conservation and storage. Around then, Congress cut the funding for his program, so he moved across the pond to Oxford to continue his pursuits. He knew Exxon had hired Whittingham to create a lithium–titanium sulfide battery. "But that effort was to fail," Goodenough says, "because dendrites form and grow across the flammable liquid electrolyte of this battery with incendiary, even explosive consequences."

Goodenough thought he had a corrective. From his earlier work, he understood that lithium–magnesium oxides were layered, so he set about exploring how much lithium he could extract from various other oxides before they became unstable. Lithium–cobalt and lithium–nickel oxides fit the bill. By 1980, his team had developed a lithium-ion battery using a lithium–cobalt oxide for the cathode, and it turned out to be a magic bullet—or at least, it allowed for a hefty charge at a lighter weight and was significantly more stable than other oxides. It's also the basic formulation you'll find inside your iPhone today. Well, almost.

Before it helped power the wireless revolution, though, the lithium-ion battery was the solution to a more mundane electronics problem. Sony was facing an obstacle to a promising new market:

camcorders. By the early 1990s, video cameras had shrunk from shoulder-mounted behemoths to handheld recorders. But the nickel-cadmium batteries used by the industry were big and bulky. "Sony needed a battery that held enough energy to run the camera but was small enough to match the camera," Sam Jaffe of Navigant Research explains. The new, ultralight rechargeable lithium-ion battery fit the bill. It wouldn't take long for the technology to spread from Sony's early Handycams to cell phones to the rest of the consumer-electronics industry.

"By the mid-1990s, almost all cameras with rechargeable batteries were using lithium-ion," Jaffe explains. "They then took over the laptop-battery market and—shortly after that—the nascent cell phone market. The same trick would be repeated in tablet computers, power tools, and handheld computing devices."

Fueled by Goodenough's research and Sony's product development, lithium batteries became a global industry unto themselves. As of 2015, they made up a thirty-billion-dollar annual market. And the trend is expected to continue, abetted by electric and hybrid vehicles. That massive, rapid-fire doubling of the market that occurred between 2015 and 2016 was primarily due to one major announcement: the opening of Tesla's Gigafactory, which is slated to become the world's largest lithium-ion-battery factory. According to Transparency Market Research, the global lithium-ion-battery market is expected to more than double to $77 billion by 2024.

★ ★ ★

It's time to hit the pool. Pools, I mean. Of lithium.

My chat with Goodenough went longer than expected, and the crew is waiting to take us to the lithium ponds that form the core of the mining operation.

"Sorry," I say to Enrique. "I was just talking with the inventor of the lithium battery."

"What did he say?" he asks, trying not to sound too interested.

"He says he's invented a better battery," I say.

"Does it use lithium?"

"No," I say. "He says it will use sodium."

"Shit."

* * *

As we drive out to the ponds through desolate desert roads, salt is in the air, underfoot, and heaped in giant piles everywhere we look. The crusted expanse and industrial machinery makes it feel a bit like an abandoned outpost. Apparently the vibe unsettles the workers too; Peña says they're a superstitious lot.

"They say they've seen Chupacabra out here," he says, "and people disappear." The harsh climate, the sprawling desert, the spare complex, the unforgiving dryness, the long salt-lined ponds—there's plenty to inspire paranormal thinking out here. I don't blame them. "And aliens. Usually, it's aliens. They say they see UFOs." Peña laughs. "Maybe they're just stopping for batteries."

We pull up to the first stop, a series of pipes stretching over white-blasted pools. SQM drills down into the brine as an oil company might drill for oil. At the Salar de Atacama, there are 319 wells pumping out 2,743 liters of that lithium-rich brine per second.

Also like an oil company, SQM is always drilling exploratory holes to locate new bounty. There are 4,075 total exploration and production boreholes, according to Peña, some of which go seven hundred to eight hundred meters deep.

The brine gets pumped into hundreds of massive evaporating ponds, where it—you guessed it—evaporates. In the high, arid desert, the process doesn't take long. Technicians blast the pipes with water twice a day to clean off encroaching salt, which can clog them up. They use the salt by-product to build everything they can—berms, tables, guardrails. I see crystals growing on a joint that was washed off just hours ago.

At the evaporation ponds, Enrique says, "You're always pumping in and pumping out." First, the workers start an evaporation route, which precipitates rock salt. Pump. Then they get potassium salt.

The lithium pools of Atacama, as photographed with my iPhone

Pump. Eventually, they concentrate the brine solution until it's about 6 percent lithium.

This vast network of clear to blue to neon-green pools is only the first step in creating the lithium that ends up in your batteries. After it's reduced to a concentrate, the lithium is shipped by tanker truck to a refinery in Salar del Carmen, by the coast.

That drive presents perhaps the most dangerous part of the process. A web of transit lines spans the area around Atacama, and the next day, Enrique, Jason, and I spend hours driving down private mining roads, passing semitrucks and tankers hauling lithium and potassium or returning to the salar for another load. More memorials marking fatal accidents dot the sides of the road. On the rare occasions it rains, flooding can shut down the entire operation and send a ripple through the entire global-supply chain. But mostly, the plight belongs to wearied drivers, taking extra trips to earn extra money and pushing the limits.

* * *

There's no spectacular white desert at Salar del Carmen, just a series of towering cylinders, a couple more pools, and rows of thrumming machinery.

The refinery operation is an industrial winter wonderland. Salt crystals grow on the reactors, and lithium flakes fall like snow on my shoulders. That's because 130 tons of lithium carbonate are whipped up here every day and shipped from Chile's ports. That's 48,000 tons of lithium a year. Because there's less than a gram of lithium in each iPhone, that's enough to make about forty-three billion iPhones.

It begins as the concentrated solution that's trucked in from Atacama and dumped into a storage pool. That's purified and then sent through a winding process of filtration, carbonization, drying, and compacting.

Soda ash is combined with the solution to form lithium carbonate, the most in-demand form of the commodity. It takes two tons of soda ash to create one ton of lithium carbonate, which is why lithium isn't refined on-site at Atacama. SQM would have to ship all that into the high desert; instead, they just bring the brine down the mountain.

As I'm walking through the flurry of lithium flakes, a hard hat on and earplugs in, past salt-jammed pipes and furiously wobbling pumps, I'm struck by the fact that much of the world's battery power originates right here. I reach out and grab a palmful, spread it around in my hands. I'm touching one part of the tangled web of a supply chain that creates the iPhone — all this just to refine a single ingredient in the iPhone's compact and complex array of technology.

From here, the lithium will be shipped from a nearby port city to a battery manufacturer, probably in China. Like most of the component parts in the iPhone, the li-ion battery is manufactured overseas. Apple doesn't make its battery suppliers public, but a host of companies, from Sony to Taipei-based Dynapack, have produced them over the years.

Even today, the type of battery that will roll off their assembly lines isn't insanely more complicated than Volta's initial formulation; the battery in the iPhone 6 Plus model, for instance, has a lithium–cobalt oxide for the cathode, graphite for the anode, and polymer for its electrolyte. It's hooked up to a tiny computer that prevents it from overheating or draining so much of its juice that it becomes unstable.

"The battery is the key to a lot of the psychology behind these devices," iFixit's Kyle Wiens points out. When the battery begins to drain too fast, people get frustrated with the whole device it powers. When the battery's in good shape, so is the phone. Predictably, the lithium-ion battery is the subject of a constant tug-of-war; as consumers, we demand more and better apps and entertainment, more video rendered in ever-higher res. Of course, we also pine for a longer-lasting battery, and the former obviously drains the latter. And Apple, meanwhile, wants to keep making thinner and thinner phones.

"If we made the iPhone a millimeter thicker," says Tony Fadell, the head of hardware for the first iPhone, "we could make it last twice as long."

* * *

About two hours after departing the world's largest lithium refinery, Jason and I got our batteries stolen. Along with the stuff they powered. We'd just left the comfy clutches of SQM; our driver had dropped us off at the bus station.

The complex looks like a dying strip mall, thick with the air of exhausted purgatory that attends midcity bus stations. I was wandering around looking for food, and Jason was watching our stuff. An old man approached him and asked him where the bus that'd just arrived was heading. While they were speaking, an accomplice strapped on my backpack and made haste for the exit. When I returned seconds later, we realized what had happened and ran frantically through the station screaming, *"¿Mochila azul?"* No dice.

We lost two laptops, our recording equipment, a backup iPhone 4s, and assorted books and notes. *But* I didn't lose this, my book, because I'd set iCloud to save my document files automatically.

It forced me to report on the rest of the trip using my iPhone alone—

voice-recording note-taking, photo-snapping—which, if I'd had just a little more data storage left, would've been entirely comfortable.

"Phone, wallet, passport," Jason took to saying as we passed through borders or left hostels, checking off our post-stolen-laptop essentials. "The only three things we have or need." It became a half-assed mantra; we'd lost a lot of valuable stuff, but we still had all the tools we needed to do everything we'd done before.

The Chilean police were friendly enough when we reported the crime but basically told us to move on—Apple products were rare and expensive in Chile, and our devices would likely be resold on the black market pronto.

* * *

Influential as li-ion has been, Goodenough believes that a new and better battery—one whose key ingredient is sodium, not lithium—is on the horizon. "We are on the verge of another battery development that will also prove societally transformational," he says. Sodium is heavier and more volatile than lithium, but cheaper and more easily accessible. "Sodium is extracted from the oceans, which are widely available, so armies and diplomacy are not required to secure the chemical energy in sodium, as is the case of the chemical energy stored in fossil fuels and in lithium," he says. There's a non-zero chance that your future iPhone battery will be powered by salt.

To which the product reviewers of the world will say: *Fine, but will our iPhone batteries ever get better? Last longer?* Whittingham thinks yes. "I think they could get double what they get today," he tells me. "The question is, would everyone be willing to pay for it?

"If you pull an iPhone apart, I think the big question we ask folks like Apple is, Do you want more efficient electronics or more energy density in your battery?" Whittingham says. They can either squeeze more juice into the battery system or tailor electronics to drain less power. To date, they've leaned primarily on the latter. For the future, who knows? "They won't give you an answer. It's a trade secret," Whittingham says.

There have been serious advances in how electronics consume

power. For one thing, "every lithium battery has total electronic protection," Whittingham notes, in the form of the computer that monitors the energy output. "They don't want you discharging it all the way," which would fry the battery.

Batteries are going to keep improving. Not just for the benefit of iPhone consumers, obviously, but for the sake of a world that's perched on the precipice of catastrophic climate change.

"The burning of fossil fuels releases carbon dioxide and other gaseous products responsible for global warming and air pollution in cities," Goodenough mentions repeatedly. "And fossil fuel is a finite resource that is not renewable. A sustainable modern society must return to harvesting its energy from sunlight and wind. Plants harvest sunlight but are needed for food. Photovoltaic cells and windmills can provide electricity without polluting the air, but this electric power must be stored, and batteries are the most convenient storage depot for electric power."

Which is exactly why entrepreneurs like Elon Musk are investing heavily in them. His Gigafactory, which will soon churn out lithium-ion batteries at a scale never before seen, is the clearest signal yet that the automotive and electronics industry have chosen their horse for the twenty-first century.

The lithium-ion battery—conceived in an Exxon lab, built into a game-changer by an industry lifer, turned into a mainstream commercial product by a Japanese camera maker, and manufactured with ingredients dredged up in the driest, hottest place on Earth—is the unheralded engine driving our future machines.

Its father just hopes that we use the powers it gives us responsibly.

"The rise of portable electronics has transformed the way we communicate with one another, and I am grateful that it empowers the poor as well as the rich and that it allows mankind to understand the metaphors and parables of different cultures," Goodenough says. "However, technology is morally neutral," he adds. "Its benefits depend on how we use it."

CHAPTER 6

Image Stabilization

A snapshot of the world's most popular camera

"Okay, here," David Luraschi whispers, furtively nodding at a man with longish greasy hair and a wrinkled leather jacket who's loping toward us on the boulevard Henri-IV. As soon as he passes, Luraschi quietly whips around, holding my iPhone vertically in front of his chest, and starts tapping the screen in rapid bursts. I, meanwhile, shove my hands in my pockets and awkwardly glance around the bustling Paris intersection. I'm trying to look inconspicuous, but feel more like a caricature of a hapless American spy.

Stalking around Paris with a professional street photographer, I felt like that pretty much all day. I've turned my iPhone over to Luraschi, who's offered to show me how he does what he does, which, I've discovered, involves a lot of waiting for an interesting subject to happen by, then tailing that person for as long as is comfortably possible.

"I walk around with the camera on and earbuds in," he says. "I use the volume buttons, here, to snap"; this has the added benefit of giving him a bit of a disguise. "Sometimes it's hard. You have to watch if you're creeping," he says, darting a quick smile at me, eyes scanning the crowd. "You don't want to creep."

We circle around the towering Bastille monument, draped in nets and scaffolding. Luraschi homes in on a woman who's dancing as she walks and snaps a beautiful photo of her, her right hand floating in the air. We pass Parisians of every stripe—stylish twentysomethings in high heels and trench coats, rumpled men with plucked-at beards, Muslim women in hijabs; Luraschi follows them all.

Luraschi, a French-American fashion photographer, had, like many artists, initially resisted the rise of Instagram and its focus on image-enhancing filters. "It's like Stephen Spielberg when he throws in a bunch of violins in a film about the Holocaust," he says. "It's like, you don't need that. The world is already pretty stylized."

But he eventually joined Instagram and surprised himself when he rose to prominence with a series of photos that shared a strong thematic link: They were all taken from behind, with the subject entirely unaware of the camera. It's harder than it looks.

The shots had a powerful symbiosis with social media—perhaps because each faceless photo could have been of just about anyone in the increasingly crowded, often anonymous public spaces of the web. Whatever the reason, the series started attracting hundreds and then thousands of shares and Likes, and before long, he was being touted as an up-and-coming phenom. People from all over the world began emailing him photos they'd taken from behind.

Instagram is of course one of the iPhone's most popular and important apps. Mashable, a website obsessed with digital culture, ranks it as the iPhone's number-one app, bar none. Released in 2010 on the heels of the suspiciously similar Hipstamatic (which pioneered the Millennial-approved photo-filter approach but wasn't free to download), it quickly developed a massive following and was scooped up by Facebook for a then-outrageous, now-bargain one billion dollars.

For Luraschi, the Instagram fame translated into more paid work, though the fear in the industry is that free amateur photos will lead

to lower salaries and less contract work for professionals. Luraschi doesn't seem concerned.

"I've always enjoyed experimenting with digital technologies," he says. "As much as I'm attached to the traditional practice of photography, of shooting on film, I like how the phone doesn't try to be a digital camera. Voyeurism has always been a big thing of photography, to not be noticed, to get access to somewhere, striking gold," he tells me.

"I've found that it being mobile and being able to fit in your pocket—it makes it easier."

* * *

Makes it easier.

If future archaeologists were to dust off advertisements for the most popular mass-market cameras of the nineteenth and twenty-first centuries, they would notice some striking similarities in the sloganeering of the two periods.

Exhibit A: You Press the Button, We Do the Rest.
Exhibit B: We've taken care of the technology. All you have to do is find something beautiful and tap the shutter button.

Exhibit A comes to us from 1888, when George Eastman, the founder of Kodak, thrust his camera into the mainstream with that simple eight-word slogan. Eastman had initially hired an ad agency to market his Kodak box camera but fired them after they returned copy he viewed as needlessly complicated. Extolling the key virtue of his product—that all a consumer had to do was snap the photos and then take the camera into a Kodak shop to get them developed—he launched one of the most famous ad campaigns of the young industry.

Exhibit B, of course, is Apple's pitch for the iPhone camera. The spirit of the two campaigns, separated by over a century, is unmistakably similar: both focus on ease of use and aim to entice the average consumer, not the photography aficionado. That principle enabled Kodak to put cameras in the hands of thousands of first-time photographers, and now it describes Apple's approach to its role as, arguably, the biggest camera company in the world.

An 1890 article in the trade magazine *Manufacturer and Builder* explained that Eastman had "the ingenious idea of combining with a camera, of such small dimensions and weight as to be readily portable, an endless strip of sensitized photographic film, so adjusted within the box of the camera, in connection with a simple feeding device, that a succession of pictures may be made—as many as a hundred—without further trouble than simply pressing a button."

Kodak's Brownie was not the first box camera; France's Le Phoebus preceded it by at least a decade. But Eastman took a maturing technology and refined it with a mass consumer market in mind. Then he promoted it. Here's Elizabeth Brayer, biographer of George Eastman: "Creating a nation (and world) of amateur photographers was now Eastman's goal, and he instinctively grasped what others in the photography industry came to realize more slowly: Advertising was the mother's milk of the amateur market. As he did in most areas of his company, Eastman handled the promotional details himself. And he had a gift for it—almost an innate ability to frame sentences into slogans, to come up with visual images that spoke directly and colorfully to everyone." Remind you of anyone?

Kodak set in motion a trend of tailoring cameras to the masses. In 1913, Oskar Barnack, an executive at a German company called Ernst Lietz, spearheaded development of a lightweight camera that could be carried outdoors, in part because he suffered from asthma and wanted to make a more easily portable option. That would become the Leica, the first mass-produced, standard-setting 35 mm camera.

* * *

In the beginning, the 2-megapixel camera that Apple tacked onto its original iPhone was hardly a pinnacle of innovation. Nor was it intended to be.

"It was more like, every other phone has a camera, so we better have one too," one senior member of the original iPhone team tells me. It's not that Apple didn't care about the camera; it's just that resources were stretched thin, and it wasn't really a priority. It certainly wasn't considered a core feature by its founder; Jobs barely mentions it in the initial keynote.

In fact, at the time of the phone's release, its camera was criticized as being subpar. Other phone makers, like Nokia, had superior camera technology integrated into their dumb phones in 2007. It would take the iPhone's growing user base—and photocentric apps like Instagram and Hipstamatic—to demonstrate to Apple the potential of the phone's camera. Today, as the smartphone market has tightened into an arms race for features, the camera has become immensely important, and immensely complex.

"There's over two hundred separate individual parts" in the

iPhone's camera module, Graham Townsend, the head of Apple's camera division, told *60 Minutes* in 2016. He said that there were currently eight hundred employees dedicated solely to improving the camera, which in the iPhone 6 is an 8-megapixel unit with a Sony sensor, optical image-stabilization module, and a proprietary image-signal processor. (Or that's one of the cameras in your iPhone, anyway, as every iPhone ships with two cameras, that one and the so-called "selfie camera.")

It's not merely a matter of better lenses. Far from it — it's about the sensors and software around them.

★ ★ ★

Brett Bilbrey was sitting in Apple's boardroom, trying to keep his head down. His boss at the time, Mike Culbert, was to his right, and the room was half full. They were waiting for a meeting to start, and everyone except Steve Jobs was seated.

"Steve was pacing back and forth and we were all trying to not catch his attention," Bilbrey says. "He was impatient because someone was late. And we're just sitting there going, *Don't notice us, don't notice us.*" Steve Jobs in a bad mood was already the stuff of legend.

Someone in the meeting had a laptop on the conference table with an iSight perched on top. Jobs stopped for a second, turned to him, looked at the external camera protruding inelegantly from the machine, and said, "That looks like shit." The iSight was one of Apple's own products, but that didn't save it from Jobs's wrath. "Steve didn't like the external iSight because he hated warts," Bilbrey says, "he hated anything that wasn't sleek and design-integrated."

Incidentally, the early iSight had been built by, among others, Tony Fadell and Andy Grignon, two men who would later become key drivers of the iPhone. The poor iSight user froze up.

"The look on his face was *I don't know what to say.* He was just paralyzed," Bilbrey says. "And without thinking at the moment, I said, 'I can fix that.'"

Well.

"Steve turned to me as if to say, *Okay, give me this revelation*. And my boss, Mike Culbert, slapped his forehead and said, 'Oh, great.'"

A new iMac was coming out, and Apple was currently switching its processor system to chips made by Intel—a huge, top secret effort that was draining the company's resources. Everyone knew that Jobs was worried that there weren't enough new features to show off, other than the new Intel chip architecture, which he feared wouldn't wow most of the public. He was on the lookout for an exciting addition to the iMac.

Jobs walked over to Bilbrey, the room dead silent, and said, "Okay, what can you do?" Bilbrey said, "Well, we could go with a CMOS imager inside and—"

"You know how to make this work?" Jobs said, cutting him off.

"Yeah," Bilbrey managed.

"Well, can you do a demonstration? In a couple weeks?" Jobs said impatiently.

"And I said, yeah, we could do it in a couple weeks. Again I hear the slap of the forehead of Mike next to me," Bilbrey tells me.

After the meeting, Culbert took Bilbrey aside. "What do you think you're doing?" he said. "If you can't do this, he's going to fire you."

★ ★ ★

This was hardly a new arena for Brett Bilbrey, but the stakes were suddenly high, and two weeks wasn't exactly a lot of time. During the 1990s, he'd founded and run a company called Intelligent Resources. Apple had hired him in 2002 to manage its media architecture group. He'd been brought on due to his extensive background in image-processing video; Bilbrey's company had made "the first video card to bridge the computer and broadcast video industries digitally," he says. Its product, the Video Explorer, "was the first computer video card to support HD video." Apple had hired him specifically because, like just about every other tech company, it had a video problem. The clumsy external camera was only part of it.

"Do you remember back in the 2001, 2002 time frame, video on the laptop was like a little window of video that was fifteen frames per second and had horrible artifacts?" Compression artifacts are what you see when you try to watch YouTube over a slow internet connection or, in ye olden times, when you'd try to watch a DVD on an old computer with a full hard drive—you'd get that dreaded picture distortion in the form of pixelated blocks. This happens when the system applies what's called lossy compression, which dumps parts of the media's data until it becomes simple enough to be stored on the available disk space (or be streamed within the bandwidth limitations, to use the YouTube example). If the compressor can't reconstruct enough data to reproduce the original video, quality tanks and you get artifacts. "The problem that we were having was, you would spend half of a video frame decoding the frame, and then the other half of the frame trying to remove as many artifacts as you could to make the picture not look like it sucked."

This was a mounting problem, as video streaming was becoming a more central part of computer use. Fixing the problem would hold the key to porting the external iSight into the hardware of a device.

"And I had this epiphany in the shower," he says. "If we don't create the blocks, we don't have to remove them. Now that sounds obvious, but how do you reconstruct video if you don't have a block?" His idea, he says, was building out an entire screen that was nothing *but* blocks. He wrote an algorithm that allowed the device to avoid de-blocking, making the entire frame of video available for playback. "So we all of a sudden were able to play full video streams on a portable Mac for that reason. One of my patents is exactly that: a de-blocking algorithm." Knowing that, he was ready to tackle the iSight issue. "Here's what I had up my sleeve: CCD imagers, which the external iSight was, were much better quality than the cheap CMOS small imagers."

There are two primary kinds of sensors used in digital cameras: charge-coupled devices, or CCDs, and complementary metal-oxide semiconductors, or CMOSs. The CCD is a light-sensitive integrated circuit that stores and displays the data for a given image so that each pixel is converted into an electrical charge. The intensity of that charge is related to a specific color on the color spectrum. In 2002, CCDs traditionally produced much better image quality but were slower and sucked down more power. CMOSs were cheaper, smaller, and allowed for faster video processing, but they were plagued with problems. Still, Bilbrey had a plan.

He would send the video from the camera down to the computer's graphics processing unit (GPU), where its extra muscle could handle color correction and clean up the video. He could offload the work of the camera sensor to the computer, basically.

So his team got to work rerouting the iSight for a demo that was now mere days away. "I developed a bunch of video algorithms for enhancement, cleanup, and filtering, and we employed many of those to create the demo," he says. One of his crack engineers started building the hardware for the unit. But the hardest part of the process wasn't the engineering—it was the politics. Building the demo meant messing with how other parts of the computer worked—and that meant messing with other teams' stuff.

"The politics of this was a nightmare," he says. "No one wanted to change the architecture that drastically. The way I got that to happen is I put it before Steve before anyone could stop me. Once Steve blesses something, no one's going to stand in the way. If you ever wanted to get something done, you just say, 'Oh, well, Steve wants to do this' and you had carte blanche, because no one was going to check with Steve to see if [he] actually said that, and no one was going to question you. So if you really wanted to win an argument in a meeting, you'd just go, 'Steve!' And then everyone would go, 'Crap.'"

With the new algorithms in place and the new hardware ready—the camera was built into the laptop lid; no more obtrusive wartcam—Bilbrey's team headed into the boardroom the night before the demonstration. They tested it, and the much more compact CMOS-powered system seemed to be flying. Seamless video in a tiny module that could fit in a laptop.

"We said, 'Okay, no one touch it—let's go home.' We're all set. It was set up in the boardroom. We'll come back tomorrow, Steve will see it, and everything's fine," Bilbrey says.

The team showed up the next day shortly before the meeting and turned the iSight on. There were two displays; the one on the left was showing the CCD, and the one on the right was showing the new-and-improved internal-ready CMOS. And, well, that image was purple. Bilbrey was flummoxed and, suddenly, terrified. "We're like, What happened?"

Just then, Jobs walked in. He looked at it and got right to the point. "The one on the right looks purple," he said.

"Yeah, we don't know what happened," Bilbrey said.

One of the software guys chimed in: "Yeah, I updated the software last night and I didn't see it then."

Bilbrey groaned. "I was like, You did what?" he says. To this day, he sounds a little incredulous. "He updated the software! I know he was just trying to do a good thing. But when everything works, you leave it alone."

Steve looked at him "with kind of this smirk" and said, simply, "Fix it. And then show it to me again." At least he wasn't going to be fired. They worked out the glitch and showed it to Jobs the next day; he signed off on it with as much brevity as he'd dismissed it the day before: "It looks great."

That was that. It was, Bilbrey says, one of the first moves in the industry toward the now-ubiquitous internal webcam.

"We got the patents for the internal camera," he says. And then, the much smaller internal iPhone cam. Bilbrey would go on to advise the engineer in charge of the first iPhone's camera, Paul Alioshin. (Alioshin, by all accounts a good and well-liked engineer, sadly passed away in a car crash in 2013.) To this day, the camera is still called the iSight. "As far as I'm aware, they're still doing it the same way. The architecture that we created to make this work is still the architecture in place."

The CMOS, meanwhile, is in the iPhone and has beaten out the CCD as the go-to technology for phone cameras today.

* * *

You can't talk about iPhone cameras without talking about selfies. FaceTime video streaming, which Bilbrey's algorithms still help de-clutter, was launched as a key feature of the iPhone 4 and would join Skype and Google Hangouts as burgeoning videoconferencing apps. Apple placed the FaceTime camera on the front side of the phone, pointed toward the user, to enable the feature, which had the added effect of making it well-designed for taking selfies.

Selfies are as old as cameras themselves (even older, if you count painted self-portraits). In 1839, Robert Cornelius was working on a new way to create daguerreotypes, the predecessor of photography. The process was much slower, so, naturally, he indulged the urge to uncover the lens, run into the shot, wait ten minutes, and replace the lens cap. He wrote on the back, *The first light picture ever taken, 1839.* The first teenager to take a photo of herself in the mirror was apparently the thirteen-year-old Russian duchess Anastasia Nikolaevna, who snapped a selfie to share with her friends in 1914. The 2000s gave rise to Myspace Photos, leading more mobile dumb phones to ship with front-facing cameras. The word *selfies* first appeared in 2002 on an Australian internet forum, but selfies really exploded with the iPhone, which, with the addition of the FaceTime cam in 2010, gave people, for better or worse, an easy way to snap photos of themselves and then filter the results.

The deluge of integrated cameras hasn't led solely to narcissistic indulgence, of course. Most of the time, people use the iSight to snap photos of their food or their babies or some striking-at-the-moment-not-so-much-at-home landscape shot. But it's also given us all the ability to document a lot more when the need comes. The immensely portable high-quality camera has given rise to citizen journalism on an unprecedented scale.

Documentation of police brutality, criminal behavior, systematic oppression, and political misconduct has ramped up in the smartphone era. Mobile video like that of Eric Garner getting choked by police, for instance, helped ignite the Black Lives Matter movement, and it has in other cases provided crucial evidence for officer wrongdoing. And protesters from Tahrir to Istanbul to Occupy have used iPhones to take video of forceful suppression, generating sympathy, support, and sometimes useful legal evidence in the process.

"Nobody even talks about that at Apple," says one Apple insider who's worked on the iPhone since the beginning. "But it's one of the things I'm most proud of being involved in. The way we can document things like that has totally changed. But I'll go into the office after Eric Garner or something like that, and nobody will ever say anything."

It goes both ways, however, and those in power can use the tool to maintain said power, as when Turkey's authoritarian leader Recep Erdoğan used FaceTime to rally supporters amid a coup.

★ ★ ★

When my wife called me, in overwhelmed ecstatic tears, to tell me that she was pregnant—a total surprise to both of us—we immediately FaceTimed. I snapped screenshots of our conversation without thinking as we tried to process this incredible development. I just did it. They're some of the most amazing images I've ever captured, full of adrenaline and love and fear and artifacts.

Before 2007, we expected to pay hundreds of dollars if we wanted to take solid digital photos. We brought digital cameras on trips, to events; we didn't expect to bring them everywhere.

Today, smartphone camera quality is close enough to the digital point-and-clicks that the iPhone is destroying large swaths of the industry. Giants like Nikon, Panasonic, and Canon are losing market share, fast. And, in a small twist of irony, Apple is using technology that those companies pioneered in order to push them out.

One of the features that routinely gets top billing in Apple's ever improving iPhone cameras is the optical image stabilization. It's a key component; without it, the ultralight iPhone, influenced by every tiny movement of your hands, would produce impossibly blurry photos.

It was developed by a man whom you've almost certainly never heard of: Dr. Mitsuaki Oshima. And thanks to Oshima, and a vacation he took to Hawaii in the 1980s, every single photo and video we've described above has come out less blurry.

A researcher with Panasonic, Oshima was working on vibrating gyroscopes for early car-navigation systems. The project ended abruptly, right before he took a fortuitous holiday in Hawaii in the summer of 1982.

"I was on vacation in Hawaii and out driving with a friend who was filming the local landscape from the car," Oshima tells me. "My friend was complaining about the difficulty of handling a huge video camera in the moving car; he couldn't stop it from shaking." Shoulder-mounted video cameras like the one his friend was using were heavy, cumbersome, and expensive, yet they still couldn't keep the picture from blurring. He made the connection between the jittery camera and his vibrating gyroscope: It occurred to him that he could eliminate blur by measuring the rotation angle of a camera with a vibrating gyro, and then correct the image accordingly. That, in the simplest terms, is what image stabilizers do today. "As soon as I returned to Japan, I started researching the possibilities for image stabilization using the gyro sensor."

Alas, his superiors at Panasonic weren't interested, and he couldn't secure a budget. But Oshima was so confident he could prove the merits of image stabilization that he took to working nights, building a prototype with a laser display-mirror device. "I remember the moment when I first turned on the prototype camera with nervous excitement. Even with a shake of the camera, the image did not blur at all. It was too good to be true! That was the most wonderful moment in my life."

Oshima chartered a helicopter to fly low around Osaka castle—one of Japan's iconic landmarks—and shot the scenery with and without his stabilization technology. The results impressed his bosses, who funded the project. Even so, after years of work, with a commercial prototype at the ready, the brass was still reluctant to bring it to market.

"There was some opposition to the commercialization of the product," Oshima says. "The Japanese market was focused on the miniaturization of the video camera, a craze that had not yet caught on in the U.S." So he turned to his counterparts in North America. "In 1988, the PV-460 video camera became the first image stabilization–equipped video camera in the world. It was a big hit in the U.S., even though it was $2,000 dollars." More expensive than the competition, sure—but the allure of steadying blurry shots proved powerful enough to warrant the extra cost.

The technology migrated to Nikon's and Canon's digital cameras in 1994 and 1995, respectively. "After that the invention immediately spread worldwide, and, consequently, my invention is employed in all digital camera image stabilizers." Over the years, he continued to work to bring the technology to more and smaller devices, and he seems awed by the ubiquity of the technology he helped pioneer.

"It is true and unbelievable that this technology is still used in almost every camera thirty-four years after the first invention," he tells me. "Now, almost every device that has a camera, including iPhones and Androids, has this image stabilization technology. My dream of equipping this technology on all cameras has finally come true."

To Oshima, innovation is the act of creating new networks between ideas—or new routes through old networks. "Inspiration is, in my understanding, a phenomenon in which one idea in the brain is stimulated to be unexpectedly associated with, and linked to, a totally different idea." It's the work of an expanded ecosystem.

Looking back through the photos Luraschi took with my phone in Paris, I'm struck by how evocative they are: an elderly woman engrossed in her book in the park; the dancing woman cutting her own path through the crowded piazza; a man standing in the open cage of a suspension bridge. Each taken quickly and seamlessly, each crystal and vivid.

My favorite shot is of a little girl climbing carelessly on an iron fence built over a retaining wall; it frames her lithe figure in an orderly web that stretches to a point on the horizon. The photo took seconds to capture and a few more to edit and share.

CHAPTER 7

Sensing Motion

From gyroscopes to GPS, the iPhone finds its place

Underneath hulking stone columns and arches, I'm standing next to Foucault's pendulum, which swings hundreds of feet down from the ceiling of this capacious, cathedral-quiet room. The pointed tip of the lead-coated bob slowly grazes a round glass table, with morning light filtering in through stained-glass windows. This is probably the closest thing to a religious experience you can have over a science experiment.

Maybe it's the stained glass. Maybe I'm still jet-lagged. Or maybe it's because the century-and-a half-old pendulum is still a little humbling to take in. But seeing it feels a bit like wandering into St. Peter's Cathedral for the first time, or peering into the Grand Canyon. There are, after all, few better ways to be viscerally reminded that you are standing on the surface of an incomprehensibly massive rock that is spinning through the void of space than staring at undeniable proof that said rock is in fact spinning.

This is the Musée des Arts et Métiers, founded in 1794, one of the oldest science and technology museums in the world. A former church abbey tucked in the middle of Paris's third arrondissement, it's at once sprawling and unassuming. A mold of the Statue of

Liberty greets visitors in the stone courtyard. You'll find some of the most important precursors to the modern computer here, from Pascal's calculator (the first automatic calculator) to the Jacquard loom (which inspired Charles Babbage to automate his Analytical Engine). And you'll find the pendulum.

Jean-Bernard-Léon Foucault—not to be confused with the more strictly philosophical Foucault, Michel—had set out to prove that the Earth rotated on its axis. In 1851, he suspended a bob and a wire from the ceiling of the Paris Observatory to show that the free-swinging pendulum would slowly change direction over the course of the day, thus demonstrating what we now call the Coriolis effect. A mass moving in a rotating system experiences a force perpendicular to the direction it's moving in and to the axis of rotation; in the Earth's Northern Hemisphere, that force deflects moving objects to the right, leading to the Coriolis effect. The experiment drew the attention of Napoléon III, who instructed him to do it again, with a bigger pendulum, at the Paris Panthéon. So Foucault built a pendulum with a wire that stretched sixty-seven meters, which impressed the emperor (and the public; Foucault's pendulum is one of the most popular exhibits at science centers around the world). The bob that Foucault used for Napoléon III swings in the Musée des Arts et Métiers today.

For his next experiment, Foucault used a gyroscope—essentially a spinning top with a structure that maintains its orientation—to more precisely demonstrate the same effect. At its fundamental level, it's not so different from the gyroscope that's lodged in your iPhone—which also relies on the Coriolis effect to keep the iPhone's screen properly oriented. It's just that today, it takes the form of a MEMS—a microelectromechanical system—crammed onto a tiny and, frankly, beautiful chip. The minuscule MEMS architectures look like blueprints for futuristic and symmetrical sci-fi temples.

The gyroscope in your phone is a vibrating structure gyroscope (VSG). It is—you guessed it—a gyroscope that uses a vibrating structure to determine the rate at which something is rotating. Here's how it

MEMS architecture

works: A vibrating object tends to continue vibrating in the same plane if, when, and as its support rotates. So the Coriolis effect—the result of the same force that causes Foucault's pendulum to rotate to the right in Paris—makes the object exert a force on its support. By measuring that force, the sensor can determine the rate of rotation. Today, the machine that does this can fit on your thumbnail. VSGs are everywhere; in addition to your iPhone, they're in cars and gaming platforms. MEMS have actually been used in cars for decades; along with accelerometers, they help determine when airbags need to be deployed.

★ ★ ★

The gyroscope is one of an array of sensors inside your phone that provide it with information about how, exactly, the device is moving through the world and how it should respond to its environment. Those sensors are behind some of the iPhone's more subtle but critical magic—they're how it knows what to do when you put it up to your ear, move it horizontally, or take it into a dark room.

To understand how the iPhone finds its place in the universe—especially in relation to you, the user—we need to take a crash course through its most crucial sensors, and two of its location-tracking chips.

When it first launched, the iPhone had only three sensors (not counting the camera sensor): an accelerometer, a proximity sensor, and an ambient-light sensor. In its debut press release, Apple extolled the virtues of one in particular.

The Accelerometer

"iPhone's built-in accelerometer detects when the user has rotated the device from portrait to landscape, then automatically changes the contents of the display accordingly," Apple's press team wrote in 2007, "with users immediately seeing the entire width of a web page, or a photo in its proper landscape aspect ratio."

That the screen would adjust depending on how you held the device was a novel effect, and Apple executed it elegantly—even if it wasn't a particularly complex technology (remember that Frank Canova had planned that feature for his canceled follow-up to the Simon, the Neon, in the 1990s).

The accelerometer is a tiny sensor that, as its name implies, measures the rate of acceleration of the device. It isn't quite as old as the gyroscope: It was initially developed in the 1920s, primarily to test the safety of aircraft and bridges. The earliest models weighed about a pound and "consisted of an E-shaped frame containing 20 to 55 carbon rings in a tension-compression Wheatstone half-bridge between the top and center section of the frame," according to industry veteran Patrick L. Walter. Those early sensors were used for "recording acceleration of an airplane catapult, passenger elevators, aircraft shock absorbers and to record vibrations of steam turbines, underground pipes and forces of explosions." The one-pound accelerometer cost $420 in 1930s bucks. For a long time, the test and evaluation community—the industry that tries to make sure our infrastructure and vehicles don't kill us—was the reason accelerometer tech continued to improve. "This group," Walter wrote, "with their aerospace and military specifications and budgets, drove the market for 50–60 years." By the 1970s, Stanford researchers developed the first MEMS accelerometers, and from the 1970s into the 2000s, the automotive industry became an important driver, using them in crash tests for airbag sensors.

And then, of course, the MEMS moved into computing. But before they'd be put in the smartphone, they'd have to make a pit stop.

"The sensors became important," says Brett Bilbrey. "Like the motion accelerometer. Do you know why it was originally put into Apple devices? It first showed up in a laptop.

"Do you remember the Star Wars light-saber app?" he asks. "People were swinging their laptops around and all that?" In 2006, two years before the idea would be turned into an app that joined the impressive cluster of mostly-useless-but-entertaining novelty apps on the iPhone, there was MacSaber—a useless-but-entertaining Mac app that took advantage of the computer's new accelerometer. It had a purpose beyond enabling laptop light-saber duels, of course; when someone knocked a laptop off a table and it went into freefall, the accelerometer would automatically shut off the hard drive to protect the data.

So Apple already had a starting point. "Then it was like, well, we want to put more sensors into the phone, we'll move the accelerometer over to the phone," Bilbrey says.

The Proximity Sensor

Let's get back to that original iPhone announcement, which noted that the "iPhone's built-in light sensor automatically adjusts the display's brightness to the appropriate level for the current ambient light, enhancing the user experience and saving power at the same time." The light sensor is pretty straightforward—a port-over from the similar laptop feature—and it's still in use today. The story behind the proximity sensor, though, is a bit more interesting.

Proximity sensors are how your iPhone knows to automatically turn off its display when you lift it to your ear and automatically revive it when you go to set it back down. They work by emitting a tiny, invisible burst of infrared radiation. That burst hits an object and is reflected back, where it's picked up by a receiver positioned next to the emitter. The receiver detects the intensity of the light. If the object—your face—is close, then it's pretty intense, and the phone knows the display should be shut off. If it receives low-intensity light, then it's okay to shine on.

"Working on that proximity sensor was really interesting," says Brian Huppi, one of the godfathers of the proto-iPhone at Apple. "It was really tricky."

Tricky because you need a proximity sensor to work for all users, regardless of what they're wearing or their hair or skin color. Dark colors absorb light, while shiny surfaces reflect it. Someone with dark hair, for instance, risks not registering on the sensor at all, and someone wearing a sequined dress might trigger it too frequently.

Huppi had to devise a hack.

"One of the engineers had really, really dark black hair and, basically, I told them, 'Go get your hair cut, get me some of your hair, and I'll glue it to this little test fixture.'" The engineer returned to work with, well, some spare hair. Huppi was true to his word: they used it to test and refine the nascent proximity sensor.

The hair almost completely absorbed light. "Light hitting this was like the worst-case scenario," Huppi says.

Even when they had the sensor up and running, it was still a precarious affair. "I remember telling one of the product designers, 'You're going to need to be really careful about how this gets mechanically implemented because it's super-sensitive,'" Huppi says. When he ran into the designer months later, he told Huppi, "Man, you were right. We had all sorts of problems with that damn thing. If there was any misalignment, the thing just didn't work."

Of course, it eventually did, and it provided another subtle touch that helped more seamlessly and gently fuse the iPhone into daily use.

GPS

Determining your phone's proximity to your head can be sussed by a sensor; determining its proximity to everything else requires a globe-spanning system of satellites. The story of why your iPhone can effortlessly direct you to the closest Starbucks begins, as so many good stories do, with the space race.

It was October 4, 1957, and the Soviets had just announced that they'd successfully launched the first artificial satellite, Sputnik 1, into orbit. The news registered as a global shock as scientists and amateur radio operators around the planet confirmed that the Russians had indeed beaten other world powers into orbit. In order to win this first leg of the space race, the Soviets had eschewed the heavy array of scientific equipment they'd initially intended to launch and instead outfitted Sputnik with a simple radio transmitter.

Thus, anyone with a shortwave radio receiver could hear the Soviet satellite as it made laps around the planet. The MIT team of astronomers assigned to observe Sputnik noticed that the frequency of radio signals it emitted increased as it drew nearer to them and decreased as it drifted away. This was due to the Doppler effect, and they realized that they could track the satellite's position by measuring its radio frequency—and also that it could be used to monitor theirs.

It took the U.S. Navy only two years from that point to develop the world's first satellite navigation system, Transit. In the 1960s and 1970s, the U.S. Naval Research Laboratory worked in earnest to establish the Global Positioning System.

Geolocation has come a long way since the space race, of course, and now the iPhone is able to get a very precise read on your whereabouts—and on your personal motion, movements, and physical activities. Today, every iPhone ships with a dedicated GPS chip that trilaterates with Wi-Fi signals and cell towers to give an accurate reading of your location. It also reads GLONAS, Russia's Cold War–era answer to GPS.

The best-known product of this technology is Google Maps. It remains the most popular mapping application worldwide and it may be the most popular map ever made, period. It has essentially replaced every other previous conception of the word *map*.

But Google Maps did not, in fact, originate with Google. It began as a project headed up by Lars and Jens Rasmussen, two Danish-born brothers who were both laid off from start-ups in the wake of the first dot-com bubble-burst. Lars Rasmussen, who describes himself as having the "least developed sense of direction" of anyone he knows, says that his brother came up with the idea after he moved home to Denmark to live with his mother.

The two brothers ended up leading a company called Where2 in Sydney, Australia, in 2004. After years of failing to interest anyone in the technology — people kept telling them they just couldn't monetize maps — they eventually sold the thing to Google, where it would be transfigured into an app for the first iPhone.

It was, probably, the iPhone's first killer app.

As the tech website the Verge noted, "Google Maps was shockingly better on the iPhone than it had been on any other platform." Pinch-to-zoom simply made navigating feel fluid and intuitive. When I asked the iPhone's architects what they thought its first must-use function was, Google Maps was probably the most frequent answer. And it was a fairly last-minute adoption; it took two iPhone software engineers, who had access to Google's data as part of that long-forgotten early partnership, about three weeks to create the app that would forever change people's relationship to navigating the world.

Magnetometer

Finally, of the iPhone's location sensors, there's the magnetometer. It has the longest and most storied history of all—because it's a compass basically. And compasses can be traced back at least as far as the Han Dynasty, around 206 B.C.

Now, the magnetometer, accelerometer, and gyroscope all feed their data into one of Apple's newer chips: the motion coprocessor, a tiny chip that the website iMore describes as Robin to the main processor's Batman. It's an untiring little sidekick, computing all the location data so the iPhone's brain doesn't have to, saving time, energy, and power. The iPhone 6 chip is manufactured by the Dutch company NXP Semiconductors (formerly Philips), and it's a key component in so-called wearable functionalities; it tracks your daily footsteps, travel distances, and elevation changes. It's the iPhone's internal FitBit — it knows whether you're riding your bike, walking, running, or driving. And it could, eventually, know a lot more.

"Over the long term, the chip could help advance gesture-recognition apps and sophisticated ways for your smartphone to anticipate your needs, or even your mental state," writes MIT's David Talbot. Whipping your phone around? It might know you're angry. And the accelerometer can already interpret a shake as an input mechanism ("shake to undo," or "shake to shuffle"). Who knows what else our black rectangles might learn about us by interpreting our minor movements and major motions.

These features are not altogether uncontroversial, however, mostly because they enable constant location tracking and technically can never be turned off. Take the story of Canadian programmer Arman Amin, who inadvertently made waves when he posted a story about traveling with his iPhone to Reddit shortly after the M chips started showing up in the 5s.

"While traveling abroad, my iPhone cable stopped working so my 5s died completely," Amin wrote. "I frequently use Argus [a fitness app] to track my steps...since it takes advantage of the M7 chip built into the phone. Once I got back from my vacation and charged the phone, I was surprised to see that Argus displayed a number of steps for the 4 days that my phone was dead. I'm both incredibly impressed and slightly terrified."

Even after Amin's battery was dead, it appears to have continued dripping power to the super-efficient M7 chip. It was a stark demonstration of a common fear that has accompanied the rise of the iPhone, and the smartphone in general—that our devices are tracking our every move. It was a reminder that even with your phone off, even with the battery dead, a chip is tracking your steps. And it feeds into concerns into raised by the iPhone's location services too, a setting that, unless disabled, regularly sends Apple data about your whereabouts.

The motion tracker helps illustrate a defining paradox of the smartphone zeitgeist: we demand always-on convenience but fear always-on surveillance. This suite of technologies, from GPS to accelerometer to motion tracker, has all but eliminated paper maps and rendered giving directions a dying art form. Yet our very physicality—our movements, migrations, relationships to the spatial world—is being uncovered, decoded, and put to use.

Over a century ago, scientists like Foucault built devices to help humanity understand the nature of our location in the universe—those devices still draw crowds of observers, myself among them, who bask in the grand, nineteenth-century demonstration of planetary motion sensing. As the old pendulum swings, the physics it proved out is working to the determine the location of the devices in our pockets.

And that science is still advancing.

"For many years my group looked at expanding the sensors of the phone," Bilbrey says, referring to Apple's Advanced Technology Group. "There are still more sensors and things I shouldn't talk about. . . . We're going to see more sensors showing up."

CHAPTER 8

Strong-ARMed

How the iPhone grew its brain

"You want to see some *old* media?"

Alan Kay grins beneath his gray mustache and leads me through his Brentwood home. It's a nice place, with a tennis court out back, but given the upper-crust Los Angeles neighborhood it sits in, it's hardly ostentatious. He shares it with his wife, Bonnie MacBird, the author and actress who penned the original script for *Tron*.

Kay is one of the forefathers of personal computing; he's what you can safely call a living legend. He directed a research team at the also-legendary Xerox PARC, where he led the development of the influential programming language Smalltalk, which paved the way for the first graphical user interfaces. He was one of the earliest advocates, back in the days of hulking gray mainframes, for using the computer as a dynamic instrument of learning and creativity. It took imagination like his to drive the computer into the public's hands.

The finest distillation of that imagination was the Dynabook, one of the most enduring conceptual artifacts of Silicon Valley—a handheld computer that was powerful, dynamic, and easy enough to operate that children could use it, not only to learn but to create media and write their own applications. In 1977, Kay and his colleague Adele

Goldberg published *Personal Dynamic Media*, and described how they hoped it would operate.

"Imagine having your own self-contained knowledge manipulator," they directed the reader—note the language and the emphasis on knowledge. "Suppose it had enough power to outrace your senses of sight and hearing, enough capacity to store for later retrieval thousands of page-equivalents of reference materials, poems, letters, recipes, records, drawings, animations, musical scores, waveforms, dynamic simulations, and anything else you would like to remember and change."

Some of the Dynabook's specs should sound familiar. "There should be no discernible pause between cause and effect. One of the metaphors we used when designing such a system was that of a musical instrument, such as a flute, which is owned by its user and responds instantly and consistently to its owner's wishes," they wrote.

The Dynabook, which looks like an iPad with a hard keyboard, was one of the first mobile-computer concepts ever put forward, and perhaps the most influential. It has since earned the dubious distinction of being the most famous computer that never got built.

I'd headed to Kay's home to ask the godfather of the mobile computer how the iPhone and a world where two billion people owned smartphones compared to what he had envisioned in the 1960s and '70s.

Kay believes nothing has yet been produced—including the iPhone and the iPad—that fulfills the original aims of the Dynabook. Steve Jobs always admired Kay, who had famously told *Newsweek* in 1984 that the Mac was the "first computer worth criticizing." In the 1980s, just before he was fired from his first stint at Apple, Jobs had been pushing an effort to get the Dynabook built. Jobs and Kay talked on the phone every couple of months until Steve's passing, and Jobs invited him to the unveiling of the iPhone in January 2007.

"He handed it to me afterwards and said, 'What do you think,

Alan—is it worth criticizing?' I told him, 'Make the screen bigger, and you'll rule the world.'"

* * *

Kay takes me into a large room. A better way to describe it might be a wing; it's a wide-open, two-story space. The first floor is devoted to a massive wood-and-steel organ, built into the far side of the wall. The second is occupied by shelves upon shelves of books; it's like a stylish public library. Old media, indeed.

We've spent the last couple of hours discussing new media, the sort that flickers by on our Dynabook-inspired devices in a barrage of links, clips, and ads. The kind that Alan Kay fears, as did his late friend the scholar and critic Neil Postman, whose book *Amusing Ourselves to Death* remains a salient critique of the modern media environment we're drowning in. In 1985 Postman argued that as television became the dominant media apparatus, it warped other pillars of society—education and politics, chiefly—to conform to standards set by entertainment.

One thrust of one of Kay's arguments—there are many—is that because the smartphone is designed as a consumer device, its features and appeal molded by marketing departments, it has become a vehicle for giving people even more of what they already wanted, a device best at simulating old media, rarely used as a knowledge manipulator.

That may be its chief innovation—supplying us with old media in new formats at a more rapid clip.

"I remembered praying Moore's law would run out with Moore's estimate," Kay says, describing the famous law put forward by computer scientist and Intel co-founder Gordon Moore. It states that every two years, the number of transistors that can fit on a square inch of microchip will double. It was based on studied observation from an industry leader and certainly wasn't a scientific law—it was, wonderfully, initially represented by Moore's slapdash hand-drawn sketch of a graph—but it became something of a self-fulfilling prophecy.

"Rather than becoming something that chronicled the progress of the industry, Moore's law became something that drove it," Moore has said of his own law, and it's true. The industry coalesced around Moore's law early on, and it's been used for planning purposes and as a way to synchronize software and hardware development ever since.

"Moore only guesstimated it for thirty years," Kay says. "So, 1995, that was a really good time, because you couldn't quite do television. Yet. Crank it up another couple of orders of magnitude, and all of a sudden everything that has already been a problem, everything that confused people, was now really cheap."

Moore's law is only now, fifty years after its conception, showing any signs of loosening its grip. It describes the reason that we can fit the equivalent of a multiple-ceiling-high supercomputer from the 1970s into a black, pocket-size rectangle today—and the reason we can stream high-resolution video seamlessly from across the world, play games with complex 3-D graphics, and store mountains of data all from our increasingly slender phones.

"If Neil were to write his book again today," Kay quips, "it would be called *Distracting Themselves to Death*."

Whether you consider the iPhone an engine of distraction, an enabler of connectivity, or both, a good place to start to understand how it's capable of each is with the transistor.

* * *

You might have heard it said that the computer in your phone is now more powerful than the one that guided the first Apollo mission to the moon. That's an understatement. Your phone's computer is way, way more powerful. Like, a hundred thousand times more powerful. And it's largely thanks to the incredible shrinking transistor.

The transistor may be the most influential invention of the past century. It's the foundation on which all electronics are built, the iPhone included; there are billions of transistors in the most modern models. When it was invented in 1947, of course, the transistor was hardly microscopic—it was made out of a small slab of a germanium, a plastic triangle, and had gold contact points that measured about half an inch. You could fit only a handful of them into today's slim iPhones.

The animating principle behind the transistor was put forward by Julius Lilienfeld in 1925, but his work lay buried in obscure journals for decades. It would be rediscovered and improved upon by scientists at Bell Labs. In 1947, John Bardeen and Walter Brattain, working under William Shockley, produced the first working transistor, forever bridging the mechanical and the digital.

Since computers are programmed to understand a binary language— a string of yes-or-no, on-or-off, or 1-or-0—humans need a way to indicate each position to the computer. Transistors can interpret our instructions to the computer; amplified could be *yes* or *on* or *1*; not amplified, *no, off, 0.*

Scientists found ways to shrink those transistors to the point that they could be etched directly into a semiconductor. Placing multiple transistors on a single flat piece of semiconducting material created an integrated circuit, or a microchip. Semiconductors—like germanium and another element you might have heard of, silicon—have unique properties that allow us to control the flow of electricity when it travels through them. Silicon is cheap and abundant (it's also called *sand*). Eventually, those silicon microchips would get a valley named after them.

More transistors mean, on a very basic level, that more complex commands can be carried out. More transistors, interestingly, do not mean more power consumption. In fact, because they are smaller, a larger number of transistors mean less energy is needed. So, to recap: As Moore's law proceeds, computer chips get smaller, more powerful, and less energy intensive.

Programmers realized they could harness the extra power to create more complex programs, and thus began the cycle you know and perhaps loathe: Every year, better devices come out that can do new and better things; they can play games with better graphics, store more high-res photos, browse the web more seamlessly, and so on.

Here's a quick timeline that should help put that into context.

The first commercial product to feature a transistor was a hearing aid manufactured by Raytheon in 1952. Transistor count: 1.

In 1954, Texas Instruments released the Regency TR-1, the first transistor radio. It would go on to start a boom that would feed the transistor industry, and it became the bestselling device in history up to that point. Transistor count: 4. So far, so good—and these aren't even on microchips yet.

Let's fast-forward, though.

The Apollo spacecraft, which landed humans on the moon in 1969, had an onboard computer, the famed Apollo Guidance Computer. Its transistors were a tangle of magnetic rope switches that had to be stitched together by hand. Total transistor count: 12,300.

In 1971, a scrappy upstart of a company named Intel released its first microchip, the 4004. Its transistors were spread over twelve square millimeters. There were ten thousand nanometers between each transistor. As the *Economist* helpfully explained, that's "about as big as a red blood cell...A child with a decent microscope could have counted the individual transistors of the 4004." Transistor count: 2,300.

The first iPhone processor, a custom chip designed by Apple and Samsung and manufactured by the latter, was released in 2007. Transistor count: 137,500,000.

That sounds like a lot, but the iPhone 7, released nine years after the first iPhone, has roughly 240 times as many transistors. Total count: 3.3 billion.

That's why the most recent app you downloaded has more computing power than the first moon mission.

Today, Moore's law is beginning to collapse because chipmakers are running up against subatomic space constraints. In the beginning of the 1970s, transistors were ten thousand nanometers apart; today, it's fourteen nanometers. By 2020, they might be separated by five nanometers; beyond that, we're talking a matter of a handful of atoms. Computers may have to switch to new methods altogether, like quantum computing, if they're going to continue to get faster.

Transistors are only part of the story, though. The less-told tale is how all those transistors came to live in a chip that could fit inside a pocket-size device, provide enough muscle to run Mac-caliber software, and not drain its battery after, like, fourteen seconds.

Through the 1990s, it was assumed most computers would be plugged in and so would have a limitless supply of juice to run their microprocessors. When it came time to look for a suitable processor for a handheld device, there was really only one game in town: a British company that had stumbled, almost by accident, on a breakthrough low-power processor that would become the most popular chip architecture in the world.

* * *

Sometimes, a piece of technology is built with an explicit purpose in mind, and it accomplishes precisely that. Sometimes, a serendipitous accident leads to a surprising leap that puts an unexpected result to good use. Sometimes, both things happen.

In the early eighties, two brilliant engineers at one of Britain's fastest-rising computer companies were trying to design a brand-new chip architecture for the central processing unit (CPU) of their next desktop machine, and they had a couple of prime directives: make it powerful, and make it cheap. The slogan they lived by was "MIPS for the masses." The idea was to make a processor capable of a million instructions per second (hence, MIPS) that the public could afford. Up to that point, chips that powerful had been tailored for industry. But Sophie Wilson and Stephen Furber wanted to make a computer that capable available to everyone.

I first saw Sophie Wilson in a brief interview clip posted on You-Tube. An interviewer asks her the question that's probably put to inventors and technology pioneers more often than any other, one that, through the course of reporting this book, I'd asked more than a few times myself: How do you feel about the success of what you created? "It's pretty huge, and it's got to be a surprise. You couldn't have been thinking that in 1983 —"

"Well, clearly we were thinking that it was going to happen," Wilson cuts in, dispensing with the usual faux-humble response. "We wanted to produce a processor used by everybody." She pauses. "And so we have."

That's not hyperbole. The ARM processor that Wilson designed has become the most popular in history; ninety-five billion have been sold to date, and fifteen billion were shipped in 2015 alone. ARM chips are in everything: smartphones, computers, wristwatches, cars, coffeemakers, you name it.

Speaking of naming it, get ready for some seriously nested acronyms. Wilson's design was originally called the Acorn RISC Machine, after the company that invented it, Acorn, and RISC, which stands for *reduced instruction set computing*. RISC was a CPU design strategy pioneered by Berkeley researchers who had noticed that most computer programs weren't using most of a given processor's instruction set, yet said processor's circuitry was nonetheless burning time and energy decoding those instructions whenever it ran. Got it? RISC was basically an attempt to make a smarter, more efficient machine by tailoring a CPU to the kinds of programs it would run.

* * *

Sophie Wilson, who is transgender, was born Roger Wilson in 1957. She grew up in a DIY family of teachers.

"We grew up in a world which our parents had made," Wilson says, meaning that literally. "Dad had a workshop and lathes and drills and stuff, he built cars and boats and most of the furniture in the house. Mom did all the soft furnishings for everything, clothes, et cetera."

Wilson took to tinkering. "By the time I got to university and I wanted a hi-fi, I built one from scratch. If I wanted, say, a digital clock, I built one from scratch," she says.

She went to Cambridge, where she failed out of the math department. That turned out to be a good thing, because she switched to computer science and joined the school's newly formed Microprocessor Society. There she met Steve Furber, another inspired gearhead; he would go on to be her engineering partner on several influential projects.

By the mid-1970s, interest in personal computing was percolating in Britain, and, just as in Silicon Valley, it was attracting businessmen on top of hackers and hobbyists. Herman Hauser, an Austrian grad student who was finishing his PhD at Cambridge and casting around for an excuse not to return home to take up the family wine business, showed up at the Microprocessor Society one day.

"Herman Hauser is somebody who is chronically disorganized," Wilson says. "In the seventies, he was trying to run his life with notebooks and pocket organizers. And that was not brilliant—he wanted something electronic. He knew it would have to be low-power, so he went and found someone who knew about low-power electronics, and that was me."

She agreed to design Hauser a pocket computer. "I started working out certain diagrams for him," she says. "Then, one time, visiting him to show him the progress, I took along my whole folder which had all the designs I was doodling along with...designs for single-board small machines and big machines and all sorts of things." Hauser was piqued. "He asked me, 'Will these things work?' and I said, 'Well, of course they will.'"

Hauser founded the company that would become Acorn—named so it would appear before *Apple Computer* in listings. Along with Furber, Wilson was Acorn's star engineer. She designed the first Acorn computer from the ground up, a machine that proved popular among hobbyists. At that time, the BBC was planning a documentary series about the computer revolution and wanted to feature a new machine that it could simultaneously promote as part of a new Computer Literacy Program, an initiative to give every Briton access to a PC.

The race to secure that contract was documented in a 2009 drama, *Micro Men,* which portrays Sophie, who still went by Roger then, as a fast-talking wunderkind whose computer genius helps Acorn win the contract. After FaceTiming with Wilson, I have to say that the portrait isn't entirely inaccurate; she's sharp, witty, and broadcasts a distinct suffers-no-fools-ishness.

The BBC Micro, as the computer would come to be called, was a huge success. It quickly transformed Acorn into one of the biggest tech companies in England. Wilson and the engineers stayed restless, of course. "It was a start-up; the reward of hard work was more hard work." They set about working on the follow-up, and almost immediately ran into trouble. Specifically, they didn't like any of the existing microprocessors they had to work with. Wilson, Furber, and the other engineers felt like they'd had to sacrifice quality to ship the Micro. "The board was upside down, the power supply wasn't very good—there was no end of nastiness to it." They didn't want to have to make so many compromises again.

For the next computer, Wilson proposed they make a multiprocessor machine and leave an open slot for a second processor—that way, they'd be able to experiment until they found the right fit. Microprocessors were booming business at the time; IBM and Motorola were dominating the commercial market with high-level systems, and Berkeley and Stanford were researching RISC. Experimenting with that second slot yielded a key insight: "The complex ones that were touted as suitable for high-level languages, as so wonderful—well, the simple ones ran faster," Wilson says.

Then the first RISC research papers from Stanford, Berkeley, and IBM were declassified, introducing Wilson to new concepts. Around then, the Acorn crew took a field trip to Phoenix to visit the company that had made their previous processor. "We were expecting a large building stacked with lots of engineers," Wilson recalls. "What we found were a couple of bungalows on the outskirts of Phoenix, with two senior engineers and a bunch of school kids." Wilson had an inkling that the RISC approach was their ticket but assumed that innovating a new microchip required a massive research budget. But: "Hey, if these guys could design a microprocessor, then we could too." Acorn would design its own RISC CPU, which would put efficiency first—exactly what they needed.

"It required some luck and happenstance, the papers being published close in time to when we were visiting Phoenix," Wilson says.

"It also required Herman. Herman gave us two things that Intel and Motorola didn't give their staff: He gave us no resources and no people. So we had to build a microprocessor the simplest possible way, and that was probably the reason that we were successful."

They also had another thing that set them apart from the competition: Sophie Wilson's brain. The ARM instruction set "was largely designed in my head—every lunchtime Steve and I would walk down to the pub with Herman, talking about where we'd gotten to, what the instruction set looked like, which decisions we'd taken." That was critical in convincing their boss they could do what Berkeley and IBM were doing and build their own CPU, Wilson says, and in convincing themselves too. "We might have been timid, but [Herman] became convinced we knew what we were talking about, by listening to us."

Back then, CPUs were already more complex than most laypeople could fathom, though they were far simpler than the subatomic transistor-stuffed microchips of today. Still, it's pretty remarkable that the microprocessor design that laid the groundwork for the chip that powers the iPhone was originally designed just by thinking it through.

Curious as to what that process might look and feel like to the mere mortal computer-using public, I asked Wilson to walk me through it.

"The first step is playing fantasy instruction set," she says. "Design yourself an instruction set that you can comprehend and that does what you want it." And then you bounce the ideas off your co-conspirator. "With that going on, then Steve is trying to comprehend the instruction set's implementation. So it's no good me dreaming up instruction sets that he can't implement. It's a dynamic between the two of us, designing an instruction that's sufficiently complex to keep me as a programmer happy and sufficiently simple to keep him as a microarchitecture implementer happy. And sufficiently small to see how we can make it work and prove it."

Furber wrote the architecture in BBC Basic on the BBC Micro.

"The very first ARM was built on Acorn machines," Wilson says. "We made ARM with computers... and only simple ones at that."

The first ARM chips came back to the Acorn offices in April of 1985.

Furber had built a second processor board that plugged into the BBC computer and took the ARM processor as a satellite. He had debugged the board, but without a CPU, he couldn't be sure it was right. They booted up the machine. "It ran everything that it should," Wilson says. "We printed out pi and broke open the champagne."

But Furber soon took a break from the celebration. He knew he had to check the power consumption, because that was the key to shipping it in the cheap plastic cases that would make the computer affordable. It had to be below five watts.

He built two test points on the board to test the current—and found, oddly, that there was no current flowing at all. "This puzzled him, and everyone else, so we prodded the board, and we discovered that the main five-volt supply to the processor wasn't actually connected. There was a fault on the board. So he was trying to measure the current flowing into that five-volt supply, and there wasn't any," Wilson says.

The thing was, though, the processor was still running. Apparently with no power at all.

How? The processor was running on leakage from the circuits next to it, basically. "The low-power big thing that the ARM is most valued for today, the reason that it's on all your mobile phones, was a complete accident," Wilson says. "It was ten times lower than Steve had expected. That's really the result of not having the right sort of tools."

Wilson had designed a powerful, fully functional 32-bit processor that consumed about one-tenth of a watt of energy.

As the CPU critic Paul DeMone noted, "it compared quite favorably to the much more complicated and expensive designs, such as the Motorola 68020, which represented the state of the art." The Motorola chip had 190,000 transistors. ARM's had a mere 25,000 but used its power much more efficiently and squeezed more performance out of its lower transistor count.

Shortly after, in an effort to continue to simplify their designs, Wilson and company built the first so-called System on Chip, or SoC, "which Acorn did casually, without realizing it was a world-defining moment." The SoC basically integrates all the components of a computer into one chip — hence the name.

Today, SoCs are rampant. There's one in your iPhone, of course.

* * *

The Acorn RISC Machine was a remarkable achievement. As Acorn's fortunes wavered, ARM's promise grew. It was spun off into its own company in 1990, a joint venture between Acorn and the company it'd once tried to beat out alphabetically — Apple. CEO John Sculley wanted to use the ARM chips in Apple's first mobile device, the Newton. Over the years, as the Newton sputtered out, Apple's stake in the company declined. But ARM surged, thanks to its low-power chips and unique business model. It's that model, Wilson insists, that launched ARM into ubiquity.

"ARM became successful for a completely different reason, and that was the way the company was set up," she says. Wilson calls it the "ecosystem model": ARM designers would innovate new chips, work closely with clients with specific demands, and then license the end designs, as opposed to selling or building them in-house. Clients would buy a license giving them access to ARM's design catalog; they could order specifications, and ARM would collect a small royalty on each device sold.

In 1997, Nokia turned to ARM to build chips for its classic 6110 handset, which was the first cell phone to feature the processor. It turned out to be a massive hit, thanks in part to the more advanced user interface and long battery life its low-power chips afforded it. Oh, and it had Snake, one of the first mobile games to become a pop-culture staple. If you're old enough to remember using a cell phone at the turn of the century, you will remember playing Snake while waiting in line somewhere.

ARM's popularity grew alongside mobile devices throughout the early aughts, and it became the obvious choice for the emergent boom of smart electronics.

"This is a company that supplies all the other companies in the world but is entrusted with their secrets," Wilson says. "Where the partners know ARM will keep its word and be a solid partner. It's in everybody's interest for it to work."

So in the end, it was twin innovations — a powerful, efficient, low-power chip alongside a collaboration-centric, license-based business model — that pushed the ARM deeper into the mainstream than even Intel. Yet you've probably heard of Intel, but maybe not ARM.

Today, Wilson is the director of integrated circuits at Broadcom. When ARM was split off from Acorn, she remained a consultant. Wilson came out as transgender in 1992 and she keeps a low profile, but is nonetheless celebrated as an inspiration to women in STEM by LGBT blogs and tech magazines aware of her work. She was recognized as one of the fifteen most important women in tech by the likes of *Maximum PC*. The blog *Gender Science* named her Queer Scientist of the Month. She cuts against the stereotype of the straight male inventor that rose to prominence in the 1980s, and it's hard not to wonder whether her genius would be more widely recognized today if she better fit the Steve Jobs mold.

Even though I knew I was testing my luck, I put the same question to her that that other hapless interviewer had: How did she feel about the rise of ARM in 2016?

"I stopped being shocked at ten billion sold."

★ ★ ★

Transistors multiplying like viruses and shrinking like Alice onto integrated, low-power ARM chips. What's it mean for us?

Well, in 2007, when the iPhone launched, that ARM-architecture chip, loaded with 1.57 million transistors (and designed and manufactured by a cohort of Samsung chipmakers working closely on-site with Apple in Cupertino, but more on that later), meant something you know very well: iOS.

That powerful, efficient processor enabled an operating system that looked and felt as fluid and modern as a Mac's, stripped down though it was to run on a phone-size device.

And it meant iOS-enabled apps.

At first, it was just a few of them. In 2007, there was no App Store. What Apple made for the platform was what you got.

Though apps held the key to propelling the iPhone into popularity and transformed it into the vibrant, diverse, and seemingly infinite ecosystem that it is today, Steve Jobs was at first adamantly opposed to allowing anyone besides Apple to create apps for the iPhone. It would take a deluge of developers calling for access, a band of persistent hackers jailbreaking the walled garden, and internal pressure from engineers and executives to convince Jobs to change course.

It was, essentially, the equivalent of a public protest stirring a leader to change policy.

* * *

Steve Jobs did in fact use the phrase *killer app* in that first keynote presentation—and it's telling what he believed that killer app would be.

"We want to reinvent the phone," Jobs declared. "What's the killer app? The killer app is making calls! It's amazing how hard it is to make calls on most phones." He then went on to demonstrate how easy Apple had made it to organize contact lists, visual voicemail, and the conference-call feature.

Remember, the original iPhone, per Apple's positioning, was to be

- A wide-screen iPod with touch controls
- A phone
- An internet communicator

The revolutionary "there's an app for that" mentality was nowhere on display.

"You don't want your phone to be like a PC," Jobs told the *New*

York Times. And "you don't want your phone to be an open platform," Jobs told the tech journalist Stephen Levy the day of the iPhone launch. "You don't want it to not work because one of three apps you loaded that morning screwed it up. Cingular doesn't want to see their West Coast network go down because of some app. This thing is more like an iPod than it is a computer in that sense."

The first iPhone shipped with sixteen apps, two of which were made in collaboration with Google. The four anchor apps were laid out on the bottom: Phone, Mail, Safari, and iPod. On the home screen, you had Text, Calendar, Photos, Camera, YouTube, Stocks, Google Maps, Weather, Clock, Calculator, Notes, and Settings. There weren't any more apps available for download and users couldn't delete or even rearrange the apps. The first iPhone was a closed, static device.

The closest Jobs came to hinting at the key function that would drive the iPhone to success was his enthusiasm for its mobile internet browser, Safari. Most smartphones offered what he called "the baby internet"—access to a text-based, unappealing shadow of the multimedia-rich glories of the web at large. Safari let you surf the web for real, as he demonstrated by loading the *New York Times'* website and clicking around. But letting developers outside Apple harness the iPhone's new platform was off the menu.

"Steve gave us a really direct order that we weren't going to allow third-party developers onto our device," Andy Grignon, a senior engineer who worked on the iPhone, says. "And the primary reasons were: it's a phone above anything else. And the second we allow some knucklehead developer to write some stupid app on our device, it could crash the whole thing and we don't want to bear the liability of you not being able to call 911 because some badly written app took the whole thing down."

Jobs had an intense hatred of phones that dropped calls, which might have driven his prioritizing the phone function in those early days.

"I was around when normal phones would drop calls on Steve,"

Brett Bilbrey, who served until 2013 as the senior manager of Apple's Advanced Technology Group, recalls. "He would go from calm to really pissed off because the phone crashed or dropped his call. And he found that unacceptable. His Nokia, or whatever it was that he was using at the time, if it crashed on him, the chances were more than likely that he'd fling and smash it. I saw him throw phones. He got very upset at phones. The reason he did not want developer apps on his phone is he did not want his phone crashing."

But developers persisted in trying, even before the phone was launched. Many had been developing Mac apps for years and were eager to take a crack at the revolutionary-looking iPhone system. They aimed blog posts and social-media entreaties at Apple, asking for developer access.

So, just weeks before the release of the first iPhone, at Apple's annual developer's conference in San Francisco, Jobs announced that they would be doing apps after all—kind of. Using the Safari engine, they could write web 2.0 apps "that look exactly and behave exactly like apps on the iPhone."

John Gruber, perhaps the best-known Apple blogger and a developer himself, explained that the "message went over like a lead balloon." Indeed. "You can't bullshit developers.... If web apps—which are only accessible over a network; which don't get app icons in the iPhone home screen; which don't have any local data storage—are such a great way to write software for iPhone, then why isn't Apple using this technique for any of their own iPhone apps?" He signed off that blog post with a blunt assessment: "If all you have to offer is a shit sandwich, just say it. Don't tell us how lucky we are and that it's going to taste delicious."

Even the iPhone engineers concurred. "They kind of opened it up to web developers, saying, 'Well, they're basically apps, right?'" Grignon says. "And the developer community was like, 'You can eat a dick, we want to write real apps.'"

So, developers were irked. And by the time it debuted, at the end of June 2007, other smartphones already allowed third-party apps.

Developers did try making those web apps for the iPhone, but, as Steve might have said himself, they mostly sucked. "The thing with Steve was that nine times out of ten, he was brilliant, but one of those times he had a brain fart, and it was like, 'Who's going to tell him he's wrong?'" Bilbrey says.

The demand for real, home-screen-living, iPhone-potential-exploiting apps led enterprising hackers to break into the iOS system for no reason other than to install their own apps on it.

Hackers started jailbreaking the iPhone almost immediately. (More on that later.) Essentially, the iPhone was one of the most intuitive, powerful mobile-computing devices ever conceived—but it took an enterprising group of hackers to let consumers really use it like one. Since their exploits were regularly covered by tech blogs and even the mainstream media, it demonstrated to the public—and Apple—a thirst for third-party apps.

With demonstrable public demand, executives and engineers behind the iPhone, especially senior vice president of iOS software Scott Forstall, started pushing Jobs to allow for third-party apps. The engineers behind the operating system and the apps that shipped with the original iPhone had already cleared the way internally for a third-party app-development system. "I think we knew we'd have to do it at some point," Henri Lamiraux, vice president of iOS software engineering, tells me. But, he says, in the initial rush to ship the first iPhone, "We didn't have time to do the frameworks and make the API clean." An API, or application program interface, is a set routines, protocols, and tools for writing software applications. "It's something we are very good at, so you're very careful what you make public."

"So at the beginning we were modeling the phone after the iPod," Nitin Ganatra says. "Everything that you could do was built into the thing."

Still, it was decided early on that the functionalities—email, web browser, maps—would be developed as apps, in part because many were ported over from the Mac OS. "We had created the tools to make it so that we could make these new apps very quickly," Ganatra says. "There was an understanding internally that we don't want to make it so there's this huge amount of setup just to build the next app that Steve thinks up."

In other words, they had an API for iOS app development more or less waiting in the wings, even if it was unpolished at launch time. "In hindsight, it was awesome that we had worked that way, but very early on we were mostly just doing that out of convenience for ourselves," Ganatra says.

If months of public outcry from developers, concerted jailbreaking operations from hackers, and mounting internal pressure from Apple's own executives wasn't enough, there was one more key ingredient that Jobs and any other pro-closed-system executives surely noticed.

"The iPhone was almost a failure when it first launched," Bilbrey says. "Many people don't realize this. Internally, I saw the volume sales—when the iPhone was launched, its sales were dismal. It was considered expensive; it was not catching on. For three to six months. It was the first two quarters or something. It was not doing well."

Bilbrey says the reason was simple. "There were no apps."

"Scott Forstall was arguing with Steve and convinced him and said, Look, we've got to put developer apps on the phone. Steve didn't want to," he says. "If an app took out the phone while you were on a phone call that was unacceptable. That could not happen to an Apple phone."

"Scott Forstall said, 'Steve, I'll put the software together, we'll make sure to protect it if a crash occurs. We'll isolate it; we won't crash the phone."

In October 2007, about four months after the iPhone launched, Jobs changed course.

"Let me just say it: We want native third party applications on the iPhone, and we plan to have an SDK [software developer's kit] in developers' hands in February," Steve Jobs wrote in an announcement posted to Apple's website. "We are excited about creating a vibrant third party developer community around the iPhone and enabling hundreds of new applications for our users." (Note that even then, Jobs, along with almost everyone, was underestimating the behemoth the app economy would become.)

Even members of the original iPhone crew credit the public campaign to open Apple's walled garden for changing Jobs's mind.

"It was the obvious thing to do," Lamiraux says. "We realized very quickly we could not write all the apps that people wanted. We did YouTube because we didn't want to let Google do YouTube, but then what? Are we going to have to write every app that every company wants to do?"

Of course not. This was arguably the most important decision Apple made in the iPhone's post-launch era. And it was made because developers, hackers, engineers, and insiders pushed and pushed. It was an anti-executive decision. And there's a recent precedent—Apple succeeds when it opens up, even a little.

"The original iPod was a dud. There were not a lot sold. iPod really took off when they added support in iTunes in Windows, because most people didn't have Macs then. This allowed them to see the value of the Apple ecosystem. I've got this player, I've got a music store, I've got a Windows device that plays well. That's when iPod really took off. And you could say the same thing for the phone," Grignon says. "It got a lot of rave reviews, but where it became a cultural game-changer was after they allowed developers in. Letting them build their software. That's what it is, right? How often do you use your phone as an actual phone? Most of the time, you're sitting in line at the grocery store Tweeting about who-gives-a-fuck. You don't use it as a phone."

No, we use the apps. We use Facebook, Instagram, Snapchat, Twitter. We use Maps. We text, but often on nonnative apps like Messenger, WeChat, and WhatsApp too.

In fact, of the use cases that Apple imagined for the product—phone, iPod, internet communicator—only one truly made the iPhone transformative.

"That's how Steve Jobs used his product," says Horace Dediu, the Apple analyst. If you think about how people spend time on their mobile devices today, you're not going to think about phone calls. "The number-one job is social media today. Number two is probably entertainment. Number three is probably directions or maps or things of that nature," Dediu says. "The basic communications of emails and other things was there, but the social media did not exist at the time. It had been invented, but it hadn't been put into mobile and hadn't been transformed. Really, it was Facebook, if I may use them sort of metaphorically, that had figured out what a phone is for."

The developer kit came out late winter of 2008, and in the summer, when Apple released its first upgrade of the iPhone, the iPhone 3G, it launched, at long last, the App Store. Developers could submit apps for an internal review process, during which they would be scanned for quality, content, and bugs. If the app was approved and if it was monetized, Apple would take a 30 percent cut. And that was when the smartphone era entered the mainstream. That's when the iPhone discovered that its killer app wasn't the phone, but a store for more apps.

"When apps started showing up on the phone, that was when the sales numbers took off," Bilbrey says. "The phone all of a sudden became a phenomenon. It wasn't the internet browser. It wasn't the iPod player. It wasn't the cell phone or anything like that."

I asked him how he could be sure that apps were the reason.

"It was a one-to-one correlation. When we announced that we were going to have apps, and we started allowing them, that's when people started buying the iPhone," he says. Not only that, but it the increase was dramatic. "It was the knee of a hockey-stick spike. You saw it struggling, and then when it got to that point, it took off and started selling."

"Apps were a gold mine," Grignon says.

Grignon says one of the first successful apps they watched conquer the new ecosystem wasn't a clever new productivity app or a flashy new multitouch-based video game. It was a fart machine.

"The guy who wrote iFart made a million dollars on that fucking app. Of course we laugh at it today, but Jesus, dude, a million fucking dollars. From an app that plays fart noises." How could that possibly be? Timing helped. "Culturally, his app was the first app to get featured on late-night talk shows. They were all making fun of him, but he was laughing all the way to the bank. That's when we started to see the relevance of these things really hit the mainstream. It was like, the iPhone's cool, but, you know — fart-app dude."

In 2007, the concept of an app had yet to infiltrate popular culture; computer users were certainly familiar with software applications, but most lay users probably thought of them as programs to be installed on a hard drive via CD-ROM.

The iPhone, with its newly built-in App Store and intuitive user interface, set the conditions for a suite of simple, easy-to-program, and easy-to-use apps. It offered the opportunity to reframe how most users approached and used software programs. Maybe the app just needed a vessel to show that all manner of interactive software goofiness was available on this platform. Maybe it needed something sufficiently ridiculous to kick down the doors.

"We didn't kick them down, we blasted them down," Joel Comm, illustrious creator of iFart, tells me. "There are just so many wonderful puns."

Joel Comm had been running online businesses since the nineties, and when Apple announced its SDK in 2008, he immediately set his small outfit, InfoMedia, to work. "Our first app was not called iFart, it was iVote. It was one of the first thousand apps that came out in 2008." Alas, the civic-minded app didn't make much of an impression. "We whiteboarded so many other ideas — you know Pokémon Go? We had an idea like that for ghosts, in 2008. But we did not laugh harder than when one of our team said, 'Let's do a fart machine.' So I said, 'Let's do that.'"

When they were just about finished with the app, they saw that another fart app, Pull My Finger, had been rejected. First, they were surprised that someone had already tried their hand at a fart app (though they were convinced theirs was better). Then they were bummed: "'Apple isn't approving fart apps,'" Comm says. "'Let's shelve it, we don't want to upset the Apple gods.' So we literally sat on it and turned to some other stuff." A little time passed, though, and the team still laughed whenever someone brought up iFart. So Comm decided to go for it anyway. "It's either submit or get off the pot," he says. "They just keep coming."

Apparently Apple had a change of heart on their flatulence policy. They soon got word that "iFart and three other fart apps got approved. One of them was Pull My Finger. I immediately put out a press release. Our app was better in every way, so we raced up the charts." They priced the app at ninety-nine cents. By Christmas, he says, they had sold just short of thirty thousand units. It was about then, as Grignon mentioned, the media got hold of it. "It went to the top of the charts, number one in the world, and stayed there for just over three weeks." George Clooney declared it his favorite app. Bill Maher said, "If your phone can fart, you're part of the problem."

The success of iFart signaled an oncoming gold rush in a new digital Wild West of an economy. "I probably netted half a million dollars and way more value in publicity and PR and credibility. Two million downloads or more." All told, it was the result of about three weeks of work from a handful of people.

"It's not any stroke of genius," Comm says. "It's the perfect timing. 'Let's make this thing fart, and let's do it in an elegant way.' I think it was the novelty of it, the production of it, the storytelling around it in the media. It was not complicated; it was a sound machine." Half a million dollars in profit from making a digital whoopee cushion. It seemed silly on one level, but it was a harbinger of a new era of opportunity for developers and a new approach.

Clearly, most of the breakout early apps weren't so lowbrow — many early successes were games, like Tap Tap Revenge, Super Monkey Ball, and a Texas Hold'Em poker app, and many were truly exciting, like Pandora Radio and Shazam, which continue to be successful today.

"Of course, now, everyone's writing fart apps, but he was the original," Grignon says. "Apple had minted this new economy. And the early gold diggers won big."

★ ★ ★

This new economy, now colloquially known as the app economy, has evolved into a multibillion-dollar market segment dominated by nouveau-riche Silicon Valley companies like Uber, Facebook, Snapchat, and Airbnb. The App Store is a vast universe, housing hopeful start-ups, time-wasting games, media platforms, spam clones, old businesses, art projects, and experiments with new interfaces.

But, given the extent to which the iPhone has entered the app into the global vernacular, I thought it was worth taking stock of what, at its core, an app actually is, and what this celebrated new market segment represents. So I reached out to Adam Rothstein, a Portland-based digital archivist and media theorist who has studied early apps; he has a collection of volvelles, which are some of the earliest — they have existed for hundreds of years.

"One way to think about apps are as 'simplified interfaces for visualizing data,'" Rothstein says. "Any app, whether social media, mapping, weather, or even a game, takes large amounts of data and presents it through a small interface, with a variety of buttons or gestures for navigating and manipulating that data. A volvelle does the same. It takes data from a chart and presents it in a round, slide-rule-like interface so the user can easily view the different data relationships."

At its simplest, a volvelle is basically a paper wheel with a data set inscribed on it that's fastened to another paper wheel with its own

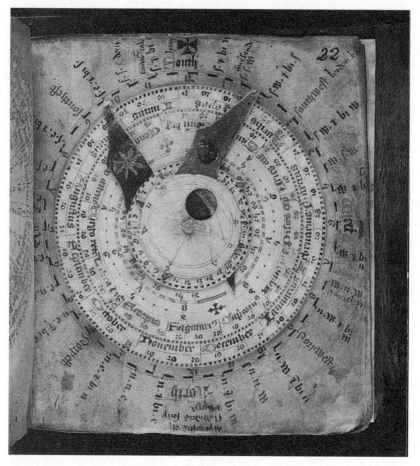

"Volvelle with three moving parts representing the zodiac, the sun, and the moon, and showing their relative positions and the moon's phases, and astronomical diagram" (British Library).

data set. The wheels can be manipulated to produce easily parsable information about the world.

They were invented by Islamic astronomers in the medieval era and were used as reference tools, navigation aids, and calculators. "They were innovative in the eleventh and twelfth century when they first appeared and relied on the relatively high technology of paper as well as the knowledge of experts and bookbinders to make them function," Rothstein says.

Some of the first mobile apps, then, were made of paper. "We've been offloading information into outsourced brains for centuries, and it is nice to feel the lineage in its continuity, from cardboard to touchscreen," Rothstein notes. Volvelles are actually considered by some historians to be primitive analog computers. "The Antikythera"—the earliest known computing device, that mysterious Greek astrolabe whose precise origins scientists have been unable to successfully pin down—"is older, and there are abacuses and counting sticks. But [the volvelle] is certainly an old-school app."

The point is that people have been using tools to simplify and wield data and coordinate solutions for centuries. Take Uber: the ride-hailing app's major innovation is its ability to efficiently pair a rider with a driver. The app reads the fluid data set of the number of available drivers in an area, taken by their GPS signals, and cross-references it with the number of desiring riders. Where those data sets intersect is where you and the driver meet for your ride. Uber is a GPS-and-Google-Maps-powered, for-profit volvelle.

"I think it's worth remembering that even as we develop new technology, we've developed many similar technologies in different forms throughout human history. We may come up with a better technology, but we are often using it to solve what is essentially the same problem. Even as we look to apply new technologies as completely and wholly new, we should remember that even as we reinvent the wheel, it's still a wheel. Many basic human needs are more timeless than we like to admit," Rothstein says.

People today may be using billions of transistors arranged on a subatomic level inside a state-of-the-art microprocessor, but a lot of the time—maybe most of the time—we're harnessing that computational power to accomplish the same sort of things as early adopters in medieval times.

★ ★ ★

That's important to keep in mind when considering the app economy.

There are, after all, over two million apps in the App Store today.

"Apple ignited the app revolution with the launch of the App Store in 2008," the company says on its website. "In just six years, the iOS ecosystem has helped create over 627,000 jobs, and U.S.-based developers have earned more than $8 billion from App Store sales worldwide."

In 2016, one report estimated that the app economy was worth $51 billion and that it would double by 2020. In early 2017, Apple announced it had paid out $20 billion to developers in 2016 and that January 1 was the single biggest day in App Store sales in the company's history; people downloaded $240 million worth of apps. Snapchat, a video-messaging app, is valued at $16 billion. Airbnb is worth $25 billion. Instagram, which was acquired by Facebook for $1 billion five years ago, is allegedly worth $35 billion now. And the biggest app-based company of all, Uber, is currently valued at $62.5 billion.

"The app industry is now bigger than Hollywood," Dediu tells me, "but nobody really talks about it." Given that Apple is such a massive company and that the vast majority of its sales come from iPhones, just how successful they have been in the App Store business can be underappreciated—remember, they take a 30 percent cut of all sales made there, just for offering the platform. "People ask, Can Apple get into becoming a services business? Well, strangely, they have done that pretty successfully, because they're making forty billion in sales on that." Dediu says Apple would be a Fortune 100 company "in services on that alone. I haven't done the numbers yet, but I think they're actually making more money than Facebook on those services. Maybe Amazon as well."

That's incredible. Apple may be making more money by hosting its App Store than two of the other biggest technology companies in the world make in total.

To better understand how the App Store might fit into a historical context, I reached out to David Edgerton, a historian of technology at Oxford University. Edgerton is the author of *The Shock of the Old,* a book that chronicles all the ways that it's usually old, persistent technologies that mold our lives. Referring to the iPhone and the app economy, he said in an email: "One of the great problems is that practically all changes in the economy in the past decades have been attributed to IT," meaning information technology. "It has become, rhetorically, sometimes the only cause of change. This is clearly absurd." He's talking about the penchant of many financial analysts and economic observers to attribute growth and progress to technology, to the iPhone and the app economy.

"One of the really massive global changes has come from the liberalization of markets, not least labour markets. There is a big difference between saying Uber is caused by IT, and saying it is caused by a desire to maximize the work of taxis and taxi drivers and have them in relentless competition with each other," he wrote, adding, "Note also that high tech once led to a world of leisure; it now leads to unrelenting work."

How powerful and transformative a force is the app economy? Has it fundamentally changed lives, or has it merely rearranged the deck chairs on a *Titanic* simulator app?

One way to find out was to get about as far away from the Silicon Valley bubble as possible, somewhere locals believe in that transformative power—a place called Silicon Savanna.

* * *

Look out the airplane window during its descent into Nairobi, and, as you close in on the city, you'll see a sprawling, unexpected stretch of savanna. Squint, and with some luck, you might be able to spot a giraffe. The airport is just a few miles outside of Nairobi National Park, the world's only such reserve inside a major metropolis. Look out that window of yours, and you'll see a bustling city in the grip of metamorphosis—skyscrapers, apartment buildings, roads, and every stripe of infrastructure under construction.

You'll have plenty of opportunity to soak in the view. Nairobi seems perpetually snarled in traffic. Banana and sugarcane vendors line the streets—at one point, after we hadn't moved for ten minutes, my driver rolls down the window and buys us a bag of sugarcane—and the city's famous matatu buses, decked out in tie-dye paint jobs or maybe a portrait of Kanye West, will inch and swerve their way past. Get closer to downtown, and you'll see a bevy of billboards and bus ads hawking mobile goods and services.

I'd traveled here to try to get a sense of just how much the app economy had affected developing economies, and I figured the country heralded as the tech savviest in Africa was a good place to look.

Kenya is uniquely mobile (in every way besides the traditional one, I guess). In 2007, the same year that the iPhone debuted, Kenya's national telecom, Safaricom, partnered with the multinational Vodafone to launch M-Pesa (*pesa* means "money" in Swahili), a mobile-payment system that allowed Kenyans to use their cell phones to easily transfer funds. Based on research that showed Kenyans had been transferring airtime among themselves as currency, not long after the system was implemented, M-Pesa took off. Since then, due to the popularity of M-Pesa, Kenya is close to becoming one of the first nations on the globe to adopt a paperless currency.

Also in 2007, Kenya saw unrest spread after the incumbent president refused to step down in the wake of a disputed election, one that independent observers deemed "flawed." Protests arose, mostly peaceful, but the police took to violently suppressing them. Hundreds of people were shot and killed. Meanwhile, violence was escalating along ethnic lines, and a crisis was soon declared.

As the reports of violent suppression began to surface, Kenya-based bloggers moved to act. Erik Hersman, Juliana Rotich, Ory Okolloh, and David Kobia built a platform called Ushahidi ("testimony" in Swahili) that allowed users to report incidents with their mobile phones. Those reports were then collated on a map so users could chart the violence, and stay safe. Ushahidi quickly caught on internationally, and it was used to track anti-immigrant violence in South Africa, to monitor elections in Mexico and India, in the aftermath of the 2010 Haiti earthquake, and during the fallout of the BP Gulf spill.

The mobile-based platforms brought Kenya to international attention as a hotbed of innovation. *Bloomberg BusinessWeek* calls Nairobi "the tech hub of Africa." TechCrunch, that great arbiter of Silicon

Valley buzz, notes, "Most discussions of the origins of Africa's tech movement circle back to Kenya." When Google sought to make inroads on the continent, it set up shop in Nairobi. *Time* magazine dubbed it "Silicon Savanna" and the name stuck. The timeline of Kenya's mobile success ran parallel to the emerging app boom, which drove mobile-minded investors, entrepreneurs, charities, and social enterprises to Nairobi.

"In your 2008 to 2012, that was the height of the mobile revolution, so everything had to have an *m*—mobile books, mobile whatever it was," Muthuri Kinyamu tells me. Kinyamu runs Communications and Programs for Nest Global and is a veteran of Nairobi's tech scene. During that time, a lot of interested groups were looking to fund mobile start-ups that could result in an Ushahidi-like success story.

In an effort to help Kenyan entrepreneurs, developers, and start-ups capitalize on the growing buzz and bring those systems to market, in 2010, Erik Hersman and his co-founders at Ushahidi founded iHub. Originally self-funded, it received a $1.4 million infusion from Omidyar Networks, the eBay founder's "philanthropic investment firm." A co-working space outfitted with high-speed internet, iHub encouraged start-ups and developers to pool skills and resources.

"It helped catalyze, helped things move faster," Hersman says. Hersman, who is American, grew up in Kenya and Sudan and has long blogged about the region on his site, the *White African*. "I think that one of the reasons that Kenya has done better in this space— Nigeria has the numbers, South Africa has the money, but the community has been tighter knit over the years. So people do tend to band together to get things done. It's a very Kenyan thing. This idea of *pamoja*, which is Swahili for 'come together.'"

As iHub is hardly a household name even in Nairobi, my Uber driver had no idea what or where it was, despite the ostensible GPS guidance from the app. We searched for it down dusty, half-paved roads before finding it on a main thoroughfare on Ngong Road. Its logo hangs high up on the fourth story of a colorful office building.

When I got inside, it was crowded and electric. Laptops on every table, coffee bar in the corner, a hum of animated conversation. Considering just the internal design, decor, and vibe alone, I could have been in Palo Alto.

I met Nelson Kwame, an entrepreneur who splits his time between his start-up, Web4All, and his freelance developer work.

Kwame was born in Sudan in 1991 after his father fled the catastrophic war that would eventually split the nation in two. He came to Kenya to attend university and is motivated by the belief that fluency in technology is the key to the region's future. He uses iHub both as a place to find potential partners and as a place to find gigs. "Let's say my friend who's a developer, his uncle works for a company that needs an app," he says, and Kwame is game. "I do a lot of websites, and a lot of apps." And he does indeed think that coding, and web development, and, increasingly, app development are crucial skills for the region's growth. His start-up organizes daylong classes in different locations to help teach coding, development, and entrepreneurship skills — he just did one in Mombasa, Kenya's major seaside port city. It was packed.

A lot of that work is coming from development groups or companies who want apps to reach an international audience. I met a young man named Kennedy Kirdi who worked for iHub's consulting arm and was building an app, funded by the UN, to help park rangers fight poachers.

And a lot of it was top-down — the U.S., investors, developers, and well-meaning charities saw the idea of the app as a handy vessel for motivating change and growth in less-developed areas. Africa had become famous for leap-frogging landlines and for its mass adoption of cell phones, so a lot of the app-boom ideology was initially copy-and-pasted onto the region, even though most Kenyans didn't have smartphones yet.

"The market share wasn't big enough for a smartphone economy to work," Eleanor Marchant, a University of Pennsylvania scholar who is embedded with iHub, tells me. "App development was faddish.

"Perception really influences how things get built," she says, "even if it wasn't an accurate perception. There was this perception of Kenya being really good at mobile. There ended up being a false equivalency," Marchant says, and a lot of early, donor-funded start-ups didn't pan out. There hadn't been another Ushahidi or M-Pesa for years. "Even, for example, the M-Pesa interface itself, it's very slow, designed to be used on an old Nokia phone, where apps don't work or don't really work. It was a text-based intervention. And it still exists that way."

In other words, the idea of a mobile revolution or an app-based revolution ported poorly from the U.S. or Europe, where it was a cultural phenomenon, to Kenya, where the reality was much different. For one thing, smartphones were simply not widely affordable yet.

"It just didn't work on the ground," Kinyamu says. I'd gone to visit Kinyamu at his new tech space, the Founder's Club, which was about to host its grand opening. A pair of monkeys ran up and down the trees in the parking lot, playfully swinging around the branches.

"We had a mix of both, say, locals, who either started up in the U.S. and came back, and we had guys from the U.S. or Europe who had come and put together a fund," he says. "And a lot of those guys actually lost their money. There were just a lot of non-African folks who were very optimistic, excited about Africa...trying out lots of things, from hackathons to system development to boot camps. There was lots of fluff. It was not the right reasons."

"It was donor-driven, so they just say, 'We're looking for something in energy,' and they're just throwing money at things," he says. "All these things, the UN, NGOs, people who decide, 'This is what Africa needs.' There was lots of that money too, and it has to be spent or deployed somehow before 2015. A lot of that was supply-driven. People looked at it more as a charity case. So if you can make—it's fluffy but it tells a better story.

"Yeah, you've won this one-million grant, but then what?" he says. "That's where the inequality comes in: There's no level playing field. If you don't have the access to the networks, to the conferences, to the ones with the money, who are quite often not based here, that's it."

Part of the problem, he says, was that investors and donors tried to import the Silicon Valley mentality to Nairobi. "Before then, it was expats lecturing guys. You know: 'This is how it works in Silicon Valley, it should work for you.'"

But Kenyans couldn't simply make a killer app and expect investors to notice it and acquire it for millions like they would in San Francisco. "People just didn't know what they were signing up for. Entrepreneurship here is more about survival. It's really hard. Start-ups are hard. But it's harder here. Because you're fixing things that the government should be able to do."

So the story of apps shifted from generalized, world-changing apps to more localized functions. "A lot of new things have in a way been localized, to put into context and apply here. It's more of, before you build programs, you understand which parts of the system you're trying to fix. You kind of have to give yourself a year of some runway. You're thinking revenue very early on. So the customer is your only source of money. And so the really solid founders you get are really doing it for food. It's about, I need to make X to pay rent, to pay my two or three guys and keep it running. I'm not doing it for the acquisition," Kinyamu says.

Thus, the newest wave of start-ups seem targeted at distinctly Kenyan interests. One app that kept coming up in conversations there was Sendy. "It's basically using the little motorcycles you see around here, turning them into delivery nodes," Hersman says. "They're moving up the chain into their Uber round." Kinyamu is trying to raise awareness about Kenya's deficient infrastructure with the Twitter-based platform What Is a Road?, which encourages users to document potholes and creates an ongoing database. And one of the fastest-rising and most successful apps of late is a way for roadside fruit vendors to coordinate with farmers. BRCK, Hersman and Rotich's latest venture, is a hardy mobile router that can give smartphones Wi-Fi in remote locations.

It helps that today, 88 percent of Kenyans have cell phones, and according to Human IPO, 67 percent of all new phones sold in Kenya are smartphones—one of the fastest adoption rates on the continent. The government has approved funds for a controversial Konza City, a fourteen-billion-dollar "techno city" that would ideally support up to 200,000 IT jobs. IT accounts for 12 percent of the economy, up from 8 percent just five years ago. Apple doesn't have much of a presence here, though the entrepreneurs I meet tell me that founders and other players wield iPhones as status symbols and brandish them about in important meetings.

And many developers seem intent on bridging profit and social entrepreneurship. "There is a sense of awakening that, hey, the binary operative of looking at start-ups as either for profit or for social entrepreneurship is flawed," Nelson Kwame says. "There is a sense that SE needs to have profit, at least for sustainability." And it goes the other way. "Beyond just making a lot of money. The whole system is moving towards this synergy." He estimates that just 30 percent of new start-ups follow the donor-funded, social-enterprise model.

Some of the core values of making apps are becoming more prominent—instead of going after grants exclusively, developers like Kwame are focusing more on what interests them. "When you build something and people use it, it's like, 'Whoa,'" he says. "After a while, there's a sense of recognition."

The story of the "mobile revolution" I found in Nairobi was anything but clear-cut; as always, it was a web of some forward-thinking innovation—by citizen bloggers and the national telecom—good marketing and storytelling, and slow progress that anointed the city as the Silicon Savanna. And while the actual impact of that designation is complex, and the app economy probably hasn't benefited most Kenyans much, the perception, both from inside and outside the iHub walls, has imbued its tech scene with a sense of drive and identity.

These are not well-off kids who trek to San Francisco to try to do an app and change the world. Most of the ones I met were extremely smart, ambitious developers (if they were in the valley of silicon instead of the savanna, they'd probably be millionaires) working, paycheck to paycheck, to bring that promised change closer to home.

★ ★ ★

Apps have, at the very least, reshaped the way people think about the delivery of software and core services. But there's another thing about the app economy: It's almost all games. In 2015, games accounted for 85 percent of the App Store's revenue, clocking in at $34.5 billion. This might not be a surprise, as some of the most popular apps are games; titles like Angry Birds or Candy Crush are impossibly ubiquitous. It's just that the app economy is often touted in its revolutionary, innovation-ushering-in capacity, not as a place to squander hours on pay-to-play puzzle games.

The App Store also became a place where lone-wolf operations can mint overnight viral successes—a Vietnamese developer, Dong Nguyen, created a simple, pixelated game called Flappy Bird that rapidly became one of the App Store's most feverishly downloaded apps. The game was difficult and now iconic; its rapid rise spawned wide discussion across the media. The game was estimated to be making fifty thousand dollars a day through the tiny ad banners displayed during play.

The other major-grossing segment of the app market besides games is subscription services. As of the beginning of 2017, Netflix, Pandora, HBO Go, Spotify, YouTube, and Hulu all ranked in the twenty top-grossing apps on the App Store. Apart from Tinder, the dating app, the rest were all games.

This too, like, a fluency with gesture-based multitouching, like the omnipresence of our cameras, like social networking, is a freshly permanent element of our lives that we can't ignore; you always have the option to dissolve your senses in a mind-obliterating app, tapping the screen and winning yourself tiny little dopamine rushes by beating levels. For all the talk about new innovative apps revolutionizing the economy, just bear in mind that when we actually put our money where our mouth is, 85 percent of the time, it's paying for distractions.

Which isn't to say there aren't great apps that can help the device serve as the knowledge manipulator Alan Kay once imagined. Or plenty of free ones that that offer worthy cultural contributions without generating much revenue. But an awful lot of the app money is going to games and streaming media—services that are engineered to be as addictive as possible.

Almost as soon as Flappy Bird rose to international fame, Nguyen decided to kill it. "Flappy Bird was designed to play in a few minutes when you are relaxed," he told *Forbes* in an interview. "But it happened to become an addictive product. I think it has become a problem." Critics couldn't fathom why he would give up a revenue stream of fifty thousand dollars a day. He had built the app himself; it was making him rich. But he was stressed and guilt-ridden. The app was too addictive. "To solve that problem, it's best to take down Flappy Bird. It's gone forever."

* * *

Which brings us back to Alan Kay. "New media just ends up simulating existing media," he says. "Moore's law works. You're going to wind up with something that is relatively inexpensive, within consumer buying habits. That can be presented as a convenience for all old media and doesn't have to show itself as new media at all," Kay says. "That's basically what's happened." Which explains why we're using our bleeding-edge new devices to stream sitcoms and play video games.

"So if look out for what the computer actually good for, what's important or critical about it, well, it's not actually that different than the printing press," he says. And that "only had to influence the autodidacts in Europe, and it didn't know where they were. So it needed to mass-produce ideas so a few percentage of the population could get them. And that was enough to transform Europe. Didn't transform most people in Europe; it transformed the movers and shakers. And what's happened."

Coincidentally, some observers have noted that the once-equal-opportunity App Store now caters more to big companies who can advertise their apps on other platforms or through already popular apps on the App Store.

And the immense, portable computing power of the iPhone is, by and large, being used for consumption. What do you use your phone for most? If you fit the profile of the average user, per Dediu, then you're using it to check and post on social media, consume entertainment, and as a navigation device, in that order. To check in with your friends, to watch videos, to navigate your surroundings. It's a stellar, routine-reordering convenience and a wonderful source of entertainment. It's a great accelerator. As Kay says, "Moore's law works." But he also laments the fact that the smartphones are designed to discourage more productive, creative interaction. "Who's going to spend serious time drawing or creating art on a six-inch screen? The angles are all wrong," he says.

What Kay is talking about is more a philosophical problem than a hardware problem. He says we have the technological capacity, the ability, to create a true Dynabook right now, but the demands of consumerism—specifically as marshaled by tech companies' marketing departments—have turned our most mobile computers into, fittingly, consumption devices. Our more powerful, more mobile computers, enabling streaming audio, video, and better graphics, have hooked us. So what could have been done differently?

"We should've included a warning label," he says with a laugh.

CHAPTER 9

Noise to Signal

How we built the ultimate network

From the top of a cell phone tower, hundreds of feet above this wide-open field, you'd swear you can see the curvature of the Earth itself, the way the horizon stretches off into the distance and gently bends out of sight. At least, that's what it looks like on YouTube — I've never been anywhere near the top of a cell tower, and I'm not about to any-time soon.

In 2008, the head of the Occupational Safety and Health Adminis-tration, Edwin Foulke, deemed tower climbing — not coal mining, highway repair, or firefighting — "the most dangerous job in America."

It's not hard to see why. The act of scaling a five-hundred-foot tower on a narrow metal ladder with a thirty-pound toolkit dangling below you on a leash with only a skimpy safety tie standing between you and a sudden plunge into terminal velocity is, well, inherently terrifying. (It's also the reason that white-knuckle videos like this — tower-climber-with-a-GoPro is a surprisingly vibrant subgenre on YouTube — rack up millions of views.) Yet people do it every day, to maintain the network that keeps our smartphones online.

After all, our iPhones would be nothing without a signal.

We only tend to feel the presence of the networks that grant us a near-constant state of connectivity to our phones when they're slow, or worse, absent altogether. Cell coverage has become so ubiquitous that we regard it as a near-universal expectation in developed places: We expect a signal much the way we expect paved roads and traffic signage. That expectation increasingly extends to wireless data, and, of course, to Wi-Fi—we've also taken to assuming homes, airports, chain coffee shops, and, public spaces will come with wireless internet access. Yet we rarely connect to the human investment it took—and still takes—to keep us all online.

As of 2016, there were 7.4 billion cell phone subscribers in a world of 7.43 billion people (along with 1.2 billion LTE wireless data subscribers). And the fact that they can call each other with relative ease is a colossal political, infrastructural, and technological achievement. For one thing, it means *a lot* of cell towers—in the United States alone, there are at least 150,000 of them. (That's an industry group estimate; good tower data is scarce as it's hard to keep track of them all.) Worldwide, there are millions.

The root of these vast networks goes back over a century, when the technology first emerged, bankrolled by nation-states and controlled by monopolies. "For the better part of the twentieth century," says Jon Agar, a historian who investigates the evolution of wireless technologies, telecoms worldwide were "organized by large single, national or state providers of the technologies," telecoms like the one run by Bell Telephone Company, started by our old friend Alexander Graham in the 1890s and that "most valuable patent"—which helped it become the largest corporation in U.S. history by the time it was broken up in 1984. "The transition from that world to our world, of many different types of individual private corporations," Agar says, is how we end up at the iPhone. It was the iPhone that "broke the carriers' backs," as Jean-Louise Gassee, a former Apple executive, once put it.

★ ★ ★

The Italian radio pioneer Guglielmo Marconi built some of the first functioning wireless-telegraphy transmitters, and he did so at the behest of a well-heeled nation-state—Britain's royal navy. A wealthy empire like that was pretty much the only entity that could afford the technology at the time. So Britain fronted the incredible costs of developing wireless communication technology, mostly in the interest of enabling their warships to keep in contact through the thick fog that routinely coated the region. There aren't many other ways to develop such large-scale, infrastructure-heavy, expensive, and difficult-to-organize networks: You need a state actor, and you need a strong justification for the network—for instance, in the U.S., police radio.

Not long after Bell Labs had announced both the invention of the transistor and cell phone technology (its creators thought the network layout looked like biological cells, hence the term "cellular") the federal government was among the first to embrace it. That's why some of the earliest radiotelephones were installed in American police cars; officers used them to communicate with police stations while they were out on patrol. You're probably still familiar with the vestiges of this system, which continues to find use, even in the age of digital communications—in what other vehicle do you still typically picture radio-dispatcher equipment?

Wireless technology remained the province of the state through most the 1950s, with one emerging exception: wealthy business folk. Top-tier mobile devices might seem expensive now, but they're not even in the same league as the first private radio-communications systems, which literally cost as much as a house. The rich didn't use radio to fight crime, of course. They used them to network their chauffeurs, allowing them to coordinate with their personal drivers, and for business.

By 1973, the networks were broad and technology advanced enough that Motorola's Martin Cooper was able to debut the first prototype mobile phone handset, famously making a public call on the toaster-size plastic cell. But the only commercially available mobile phones were car-based until the mid-1980s arrival of Motorola's DynaTAC—the series of phones Frank Canova would experiment with to create the first smartphone. These were still ultra-expensive and rare, meeting the demands of a narrow niche—the rich futurist businessman, or

Martin Sheen in *Wall Street,* pretty much. There wouldn't be a serious consumer market for mobile phones until the 1990s.

The early mobile networks that emerged to administer them were run by top-down telecoms, much like the expansive landline networks built out since the earliest days of Bell's telephone. Cell markets remained tethered to regional or nationally operated base stations, but with one exception: the more egalitarian and consumer-oriented Nordic Mobile Telephony system.

* * *

Scandinavian nations pioneered wireless before many others out of sheer necessity—routing telephone wires through vast expanses of rocky, snowy terrain was difficult. In the rest of Europe the telecom model was rigid, conventional: national systems provided by the national telecom's provider. Not so in the Nordic countries, where Swedes, Finns, Norwegians, and Danes wanted their car phones to work across borders—and the germ of roaming was planted. The Nordic Mobile Telephony system, which was established in 1981, marked the re-conception of the phone as something that could— and should—transcend borders, and reshaped the way that people thought about mobile communications, from an implement useful in local markets to a more general, more universal tool. In fact, the NMT network's stated goal was building a system in which "everyone could call everyone." It used a computerized register to keep tabs on people's locations while they were roaming. It would become the first automatic mobile network and a model for all advanced wireless networks to follow.

"It has certain features which are really influential," Agar says. "The shared Scandinavian values towards design, with open attitudes towards technologies. One of the crucial things was a willingness to set aside some of those purely national interests in favor of something that was more consumer-oriented. That was about, for example, roaming between countries." Unsurprisingly, the more open, unbound system proved popular. So popular, in fact, that it effectively minted the model for the mobile standard that would go on to conquer the world.

In 1982, European telecom engineers and administrators met under the banner of the Groupe Spécial Mobile to hash out the future of the continent's cell system, and to discuss whether a unified cell standard was technically and politically possible. See, the European Commission wanted to do for all of Europe what NMT had done for the Nordic nations. It's rarely sexy to discuss vast, slow-moving bureaucracies, but GSM is an incredible triumph of political collaboration. Though it would take a decade to orchestrate the GSM test program, complete the technical specifications, and align the politics, it was a behemoth effort of technical cooperation and diplomatic negotiation. To drastically oversimplify matters, there were those who sought a stronger, more united Europe, and those who argued its states should be more independent; GSM was seen as a vessel for uniting Europe, and thus championed by the European Commission. "The best illustration of GSM as a politically charged European project is given by the facility to roam," Agar says. "Just as in NMT, roaming, the ability to use the same terminal under different networks, was prioritized, even though it was expensive, because it demonstrated political unity." If citizens of different European nations could call each other on the go or easily phone home when abroad, the constellation of diverse countries might feel a bit more like they were part of the same neighborhood.

When it finally launched in 1992, GSM would cover eight EU nations. Within three years, it covered almost all of Europe. Renamed the Global System for Mobile, it soon lived up to its moniker. By the end of 1996, GSM was used in 103 countries, including the United States, though it often wasn't the only available standard. Today, it's pretty much everywhere—an estimated 90 percent of all cell calls in 213 countries are made over GSM networks. (The U.S. market is one of the few that's split; Verizon and Sprint use a competing standard, called CDMA, while T-Mobile and AT&T use GSM. One easy way to tell if your phone is GSM: it comes with an easily removable subscriber identity module—a SIM card.)

Without the EU's drive to standardize mobile—and its push for unity—we might not have seen such wide-scale and rapid adoption of cell phones. Critics have railed against some of GSM's specifications as being overly complicated. It's been called the "Great Software Monster" and the "most complicated system built by man since the Tower of Babel," but maybe that's fitting—standardizing network access for much of the globe, allowing "everyone to talk to everyone"—was a crucial feat.

★ ★ ★

While wireless cell networks evolved from massive government-backed projects, the main way our phones get online began as a far-flung academic hackaround. Wi-Fi began long before the web as we know it existed and was actually developed along the same timeline as ARPANET. The genesis of wireless internet harkens back to 1968 at the University of Hawaii, where a professor named Norman Abramson had a logistics problem. The university had only one computer, on the main campus in Honolulu, but he had students and colleagues spread across departments and research stations on the other islands. At the time, the internet was routed through Ethernet cables—and routing an Ethernet cable hundreds of miles underwater to link the stations wasn't an option.

Not entirely unlike the way harsh northern terrain drove the Scandinavians to go wireless, the sprawling expanse of Pacific Ocean forced Abramson to get creative. His team's idea was to use radio communications to send data from the terminals on the small islands to the computer on Honolulu, and vice versa. The project would grow into the aptly named ALOHAnet, the precursor to Wi-Fi. (One of those reverse-engineered acronyms, it originally stood for Additive Links On-line Hawaii Area.) The ARPANET is the network that would mutate into the internet, and it's fair to say that ALOHAnet would do the same with Wi-Fi.

At the time, the only way to remotely access a large information processing system—a computer—was through a wire, via leased lines or dial-up telephone connections. "The goal of THE ALOHA

SYSTEM is to provide another alternative for the system designer and to determine those situations where radio communications are preferable to conventional wire communications," Abramson wrote in a 1970 paper describing the early progress. It's a frank, solutions oriented declaration that—like E. A. Johnson's touchscreen patent— downplays (or underestimates) the potential of the innovation it describes.

The way that most radio operators went about sharing an available channel—a scarce resource in a place like Hawaii, where there's less network coverage—was by dividing it up into either time slots or frequency bands, then giving each wayward station one of either. Once each party gets a band of frequency or a time slot—and only then—they could open communication.

On Hawaii, though, with the university's slow mainframe, that meant dragging data transfer to a crawl. Thus, ALOHAnet's chief innovation: It would be designed with only two high-speed UHF channels, one uplink, one downlink. The full channel capacity would be open to all, which meant that if two people tried to use it at once, a transmission could fail. In which case, they'd just have to try again. This system would come to be known as random access protocols. ARPANET nodes could communicate directly only with a node at the other end of a wire (or satellite circuit). "Unlike the ARPANET where each node could only talk directly to a node at the other end of a wire or satellite circuit, in ALOHAnet all client nodes communicated with the hub on the same frequency."

In 1985, the FCC opened the industrial, scientific, and medical (ISM) band for unlicensed use, allowing interested parties to do exactly that. A group of tech companies convened around a standard in the 90s, and marketing flacks gave it the entirely meaningless name "wireless fidelity"—concocted to sound like Hi-Fi—and Wi-Fi was born.

* * *

As GSM grew in Europe and around the world and as the cost of mobile phones fell, more users inevitably got their hands on the technology. And, as Agar says, "The key uses of a technology are discovered by users. They're not necessarily at the fore of the mind of the designers." So, it didn't take long for those users to discover a feature, added as an afterthought, that would evolve into the cornerstone of how we use phones today. Which is why we should thank Norwegian teens for popularizing texting.

A researcher named Friedhelm Hillebrand, the chairman of GSM's nonvoice services committee, had been carrying out informal experiments on message length at his home in Bonn, Germany. He counted the characters in most messages he tapped out and landed on 160 as the magic number for text length. "This is perfectly sufficient," he thought. In 1986, he pushed through a requirement mandating that phones on the network had to include something called short message service, or SMS. He then shoehorned SMSing into a secondary data line originally used to send users brief updates about network status.

Its creators thought that text messaging would be useful for an engineer who was, say, out in the field checking on faulty wires— they'd be able to send a message back to the base. It was almost like a maintenance feature, Agar says. But it also enabled text messaging to appear on most phones. It was a minuscule part of the sprawling GSM, and engineers barely texted. But teenagers, who were amenable to a quick, surreptitious way to send messages, discovered the service. Norwegian teenagers, Agar says, took to texting in far greater numbers than any network engineers ever did. During the nineties, texting was largely a communication channel for youth culture.

This principle plays out again and again throughout the history of technology: Designers, marketers, or corporations create a product or service, users decide what they actually want to do with it. This happened in Japan at the turn of the century too: The telecom NTT DoCoMo had built a subscription service for mobile internet aimed at businessmen called i-Mode. The telecom tightly curated the websites that could appear on the screen, and pitched services like airline ticket reservations and email. It flopped with business class, but was taken up by twentysomethings, who helped smartphones explode in Japan nearly a decade before they'd do the same in the U.S.

The user takeover phenomenon has happened a number of times on the iPhone as well; Steve Jobs, recall, said that the iPhone's killer app was making calls. (To his credit, phone-focused features like visual voicemail *were* major improvements.) Third-party apps weren't allowed. Yet users eventually dictated that apps would be central and that making calls would be a bit be less of a priority.

As everyone began talking to everyone, as the teens started texting, and as we all hit the App Store, wireless networks stretched out across the globe. And somebody had to build, service, and repair the countless towers that kept everyone connected.

* * *

In the summer of 2014, Joel Metz, a cell-tower climber and twenty-eight-year-old father of four, was working on a tower in Kentucky, 240 feet above the ground. He was replacing an old boom with a new one when his colleagues heard a loud pop; a cable suddenly flew loose and severed Metz's head and right arm, leaving his body dangling hundreds of feet in the air for six hours.

The gruesome tragedy is, sadly, not a fluke. Which is why it's absolutely necessary to interrupt the regularly scheduled story of collaboration, progress, and innovation with a reminder that it all comes at a cost, that the infrastructure that makes wireless technology possible is physically built out by human labor and high-risk work, and that people have died to grow and maintain that network. Too many people. Metz's death was just one of scores that have befallen cell-tower climbers over the past decade.

Since 2003, Wireless Estimator, the primary industry portal for tower design, construction, and repair, has tallied 130 such accidental deaths on towers. In 2012, PBS *Frontline* and ProPublica partnered for an investigation into the alarming trend. An analysis of Occupational Safety and Health Administration (OSHA) records showed that tower climbing had a death rate that was ten times that of the construction industry's. The investigators found that climbers were often given inadequate training and faulty safety equipment before being sent out to do maintenance work on structures that loomed hundreds of feet above the ground. If you ever want to get a taste of how gut-churningly high these workers climb, there's always YouTube; watch, and get vicarious vertigo, on the LTE network they help keep running.

The investigation found that one carrier was seeing more fatalities than all of its major competitors combined. Guess who, and guess when: "Fifteen climbers died on jobs for AT&T since 2003. Over the same period, five climbers died on T-Mobile jobs, two died on Verizon jobs and one died on a job for Sprint," the report noted. "The death toll peaked between 2006 and 2008, as AT&T merged its network with Cingular's and scrambled to handle traffic generated by the iPhone." Eleven climbers were killed during that rush.

You might recall the complaints about AT&T's network that poured in after the iPhone debuted; it was soon overloaded, and Steve Jobs was reportedly furious. AT&T's subsequent rush to build out more tower infrastructure for better coverage, ProPublica's report indicated, contributed to hazardous working conditions and the higher-than-usual death toll.

The following years saw fewer deaths, down to a low of just one in 2012. Sadly, after that, there was another sharp spike—up to fourteen deaths in 2013. The next year, the U.S. Labor Department warned of "an alarming increase in worker deaths." The major carriers typically offload tower construction and maintenance to third-party subcontractors, who often have less-than-stellar safety records. "Tower worker deaths cannot be the price we pay for increased wireless communication," OSHA's David Michaels said in a statement.

Tower climbing is known as a high-risk, high-reward job. Ex-climbers have described it as a "Wild West environment," and a fraction of those who've died in accidents have tested positive for alcohol and drugs. Still, the subcontractors are rarely significantly penalized when deaths do occur, and with no substantial reduction in the rate of death, we have to assume that until regulators clamp down or the rate of expansion slows, the loss of lives will persist.

We need to integrate this risk, this loss, into our view of how technology works. We might not have developed wireless radio communications without Marconi, cell phones without Bell Labs, a standardized network without EU advocates—and we wouldn't get reception without the sacrifice of workers like Joel Metz. Our iPhones wouldn't have a network to run on without all of the above.

These forces combined have propelled a vast expansion of smartphones: There were 3.5 million smartphone subscribers in the U.S. in 2005—and 198 million in 2016. That's the gravitational power of the iPhone in action; it reaches back into the networks of the past and ripples out into the drive to build the towers of the future.

iii: Enter the iPhone

Slide to unlock

If you worked at Apple in the mid-2000s, you might have noticed a strange phenomenon afoot. People were disappearing. It happened slowly at first. One day there'd be an empty chair where a star engineer used to sit. A key member of the team, gone. Nobody could tell you exactly where they went.

"I had been hearing rumblings about, well, it was unclear what was being built, but it was clear that a lot of the best engineers from the best teams had been slurped over to this mysterious team," says Evan Doll, who was a software engineer at Apple then.

Here's what was happening to those star engineers. First, a couple of managers had shown up in their office unannounced and closed the door behind them. Managers like Henri Lamiraux, a director of software engineering, and Richard Williamson, a director of software.

One such star engineer was Andre Boule. He'd been at the company only a few months.

"Henri and I walked into his office," Williamson recalls, "and we said, 'Andre, you don't really know us, but we've heard a lot about you, and we know you're a brilliant engineer, and we want you to come work with us on a project we can't tell you about. And we want you to do it now. Today.'"

Boule was incredulous, then suspicious. "Andre said, 'Can I have

197

some time to think about it?'" Williamson says. "And we said, 'No.'" They wouldn't, and couldn't, give him any more details. Still, by the end of the day, Boule had signed on. "We did that again and again across the company," Williamson says. Some engineers who liked their jobs just fine said no, and they stayed in Cupertino. Those who said yes, like Boule, went to work on the iPhone.

And their lives would never be the same—at least, not for the next two and a half years. Not only would they be working overtime to hammer together the most influential piece of consumer technology of their generation, but they'd be doing little else. Their personal lives would disappear, and they wouldn't be able to talk about what they were working on. Steve Jobs "didn't want anyone to leak it if they left the company," says Tony Fadell, one of the top Apple executives who helped build the iPhone. "He didn't want anyone to say anything. He just didn't want—he was just naturally paranoid."

Jobs told Scott Forstall, who would become the head of the iPhone software division, that even he couldn't breathe a word about the phone to anyone, inside Apple or out, who wasn't on the team. "He didn't want, for secrecy reasons, for me to hire anyone outside of Apple to work on the user interface," Forstall said. "But he told me I could move anyone in the company into this team." So he dispatched managers like Henri and Richard to find the best candidates. And he made sure potential recruits knew the stakes up-front. "We're starting a new project," he told them. "It's so secret, I can't even tell you what that new project is. I cannot tell you who you will work for. What I can tell you is if you choose to accept this role, you're going to work harder than you ever have in your entire life. You're going to have to give up nights and weekends probably for a couple years as we make this product."

And "amazingly," as Forstall put it, some of the top talent at the company signed on. "Honestly, everyone there was brilliant," Williamson tells me. That team—veteran designers, rising programmers, managers who'd worked with Jobs for years, engineers who'd never met him—would end up becoming one of the great, unheralded creative forces of the twenty-first century.

One of Apple's greatest strengths is that it makes its technology look and feel easy to use. There was nothing easy about making the iPhone, though its inventors say the process was often exhilarating.

Forstall's prediction to the iPhone team would be borne out.

"The iPhone is the reason I'm divorced," Andy Grignon, a senior iPhone engineer, tells me. I heard that sentiment more than once throughout my dozens of interviews with the iPhone's key architects and engineers. "Yeah, the iPhone ruined more than a few marriages," says another.

"It was really intense, probably professionally one of the worst times of my life," Andy Grignon says. "Because you created a pressure cooker of a bunch of really smart people with an impossible deadline, an impossible mission, and then you hear that the future of the entire company is resting on it. So it was just like this soup of misery," Grignon says. "There wasn't really time to kick your feet back on the desk and say, 'This is going to be really fucking awesome one day.' It was like, 'Holy fuck, we're fucked.' Every time you turned around there was some just imminent demise of the program just lurking around the corner."

Making the iPhone

The iPhone began as a Steve Jobs–approved project at Apple around the end of 2004. But as we've seen, its DNA began coiling long before that.

"I think a lot of people look at the form factor and they think it's not just like any other computer, but it is—it's just like any other computer," Williamson says. "In fact, it's *more* complex, in terms of software, than many other computers. The operating system on this is as sophisticated as the operating system on any modern computer. But it is an evolution of the operating system we've been developing over the last thirty years."

Like many mass-adopted, highly profitable technologies, the iPhone has a number of competing origin stories. There were as many as five different phone or phone-related projects—from tiny research endeavors to full-blown corporate partnerships—bubbling up at Apple by the middle of the 2000s. But if there's anything I've learned in my efforts to pull the iPhone apart, literally and figuratively, it's that there are rarely concrete beginnings to any particular products or technologies—they evolve from varying previous ideas and concepts and inventions and are prodded and iterated into newness by restless minds and profit motives. Even when the company's executives were under oath in a federal trial, they couldn't name just one starting place.

"There were many things that led to the development of the iPhone at Apple," Phil Schiller, senior vice president of worldwide marketing, said in 2012. "First, Apple had been known for years for being the creator of the Mac, the computer, and it was great, but it had small market share," he said. "And then we had a big hit called the iPod. It was the iPod hardware and the iTunes software. And this really changed everybody's view of Apple, both inside and outside the company. And people started asking, Well, if you can have a big hit with the iPod, what else can you do? And people were suggesting every idea, make a camera, make a car, crazy stuff."

And make a phone, of course.

Open the Pod Bay Doors

When Steve Jobs returned to take the helm of a flailing Apple in 1997, he garnered acclaim and earned a slim profit by slashing product lines and getting the Mac business back on track. But Apple didn't reemerge as a major cultural and economic force until it released the iPod, which would mark its first profitable entry into consumer electronics and become a blueprint and a springboard for the iPhone in the process.

"There would be no iPhone without the iPod," says a man who

helped build both of them. Tony Fadell, sometimes dubbed "the Podfather" by the media, was a driving force in creating Apple's first bona fide hit device in years, and he'd oversee hardware development for the iPhone. As such, there are few better people to explain the bridge between the two hit devices. We met at Brasserie Thoumieux, a swank eatery in Paris's gilded seventh arrondissement, where he was living at the time.

Fadell is a looming figure in modern Silicon Valley lore, and he's divisive in the annals of Apple. Brian Huppi and Joshua Strickon praise him for his audacious, get-it-done management style ("Don't take longer than a year to ship a product" is one of his credos) and for being one of the few people strong enough of will to stand up to Steve Jobs. Others chafe at the credit he takes for his role in bringing the iPod and iPhone to market; he's been called "Tony Baloney," and one former Apple exec advised me "not to believe a single word Tony Fadell says." After he left Apple in 2008, he co-founded Nest, a company that crafted smart home gadgets, like learning thermostats, and was acquired by Google for $3.2 billion.

Right on time, Fadell strode in; shaved head save for some stubble, icy blue eyes, snug sweater. He was once renowned for his cyberpunk style, his rebellious streak, and a fiery temper that was often compared to Jobs's. Fadell is still undeniably intense, but here, speaking easy French to the waitstaff, he was smack in the overlap of a Venn diagram showing Mannered Parisian Elite and Brash Tech Titan.

★ ★ ★

"The genesis of the iPhone, was—well, let's get started with—was iPod dominance," Fadell says. "It was fifty percent of Apple's revenue." But when iPods initially shipped in 2001, hardly anyone noticed them. "It took two years," Fadell says. "It was only made for the Mac. It was less than one percent market share in the U.S. They like to say 'low single digits.'" Consumers needed iTunes software to load and manage the songs and playlists, and that software ran only on Macs.

"Over my dead body are you gonna ship iTunes on a PC," Steve

Jobs told Fadell, he says, when Fadell pushed the idea of offering iTunes on Windows. Nonetheless, Fadell had a team secretly building out the software to make iTunes compatible with Windows. "It took two years of failing numbers before Steve finally woke up. Then we started to take off, then the music store was able to be a success." That success put iPods in the hands of hundreds of millions of people—more than had ever owned Macs. Moreover, the iPod was hip in a fashionably mainstream way that lent a patina of cool to Apple as a whole. Fadell rose in the executive ranks and oversaw the new product division.

Launched in 2001, a hit by 2003, the iPod was deemed vulnerable as early as 2004. The mobile phone was seen as a threat because it could play MP3s. "So if you could only carry one device, which one would you have to choose?" Fadell says. "And that's why the Motorola Rokr happened."

Rokring Out

In 2004, Motorola was manufacturing one of the most popular phones on the market, the ultrathin Razr flip phone. Its new CEO, Ed Zander, was friendly with Jobs, who liked the Razr's design, and the two set about exploring how Apple and Motorola might collaborate. (In 2003, Apple execs had considered buying Motorola outright but decided it'd be too expensive.) Thus the "iTunes phone" was born. Apple and Motorola partnered with the wireless carrier Cingular, and the Rokr was announced that summer.

Publicly, Jobs had been resistant to the idea of Apple making a phone. "The problem with a phone," Steve Jobs said in 2005, "is that we're not very good going through orifices to get to the end users." By *orifices*, he meant carriers like Verizon and AT&T, which had final say over which phones could access their networks. "Carriers now have gained the upper hand in terms of the power of the relationship with the handset manufacturers," he continued. "So the handset manufacturers are really getting these big thick books from the carriers telling them here's what your phone's going to be. We're not good at that." Privately, he had other reservations. One former Apple executive who had daily meetings with Jobs told me that the carrier issue wasn't his biggest hang-up. He was concerned with a lack of focus in the company, and he "wasn't convinced that smartphones were going to be for anyone but the 'pocket protector crowd,' as we used to call them."

Partnering with Motorola was an easy way to try to neutralize a threat to the iPod. Motorola would make the handset; Apple would do the iTunes software. "It was, How can we make it a very small experience, so they still had to buy an iPod? Give them a taste of iTunes and basically turn it into an iPod Shuffle so that they'll want to upgrade to an iPod. That was the initial strategy," Fadell says. "It was, 'Let's not cannibalize the iPod because it's going so well.'"

As soon as the collaboration was made public, Apple's voracious rumor mill started churning. With an iTunes phone on the horizon, blogs began feeding the anticipation for a transformative mobile device that had been growing for some time already.

Inside Apple, however, expectations for the Rokr could not have been lower. "We all knew how bad it was," Fadell says. "They're slow, they can't get things to change, they're going to limit the songs." Fadell laughs aloud when discussing the Rokr today. "All of these things were coming together to make sure it was really a shitty experience."

But there may have been another reason that Apple's executives were tolerating the Rokr's unfurling shittiness. "Steve was gathering information during those meetings" with Motorola and Cingular, Richard Williamson says. He was trying to figure out how he might pursue a deal that would let Apple retain control over the design of its phone. He considered having Apple buy its own bandwidth and become its own mobile virtual network operator, or MVNO. An executive at Cingular, meanwhile, began to cobble together an alternative deal Jobs might actually embrace: *Give Cingular exclusivity, and we'll give you complete freedom over the device.*

Fix What You Hate

From Steve Jobs to Jony Ive to Tony Fadell to Apple's engineers, designers, and managers, there's one part of the iPhone mythology that everyone tends to agree on: Before the iPhone, everyone at Apple thought cell phones "sucked." They were "terrible." Just "pieces of junk." We've already seen how Jobs felt about phones that dropped calls.

"Apple is best when it's fixing the things that people hate," Greg Christie tells me. Before the iPod, nobody could figure out how to use a digital music player; as Napster boomed, people took to carting around skip-happy portable CD players loaded with burned albums. And before the Apple II, computers were mostly considered too complex and unwieldy for the layperson.

"For at least a year before starting on what would become the iPhone project, even internally at Apple, we were grumbling about how all of these phones out there were all terrible," says Nitin Ganatra, who managed Apple's email team before working on the iPhone. It was water-cooler talk. But it reflected a growing sense inside the company that since Apple had successfully fixed—transformed, then dominated—one major product category, it could do the same with another.

"At the time," Ganatra says, "it was like, 'Oh my God, we need to go in and clean up this market too—why isn't Apple making a phone?'"

Calling All Pods

Andy Grignon was restless. The versatile engineer had been at Apple for a few years, working in different departments on various projects. He's a gleefully imposing figure—shaved-head bald, cheerful, and built like a friendly bear. He had a hand in everything from creating the software that powered the iPod to working on the software for a videoconferencing program and iChat. He'd become friends with rising star Tony Fadell when they'd built the iSight camera together.

After wrapping up another major project—writing the Mac feature Dashboard, which Grignon affectionately calls "his baby" (it's the widget-filled screen with the calculator and the calendar and so on)— he was looking for something fresh to do. "Fadell reached out and said, 'Do you want to come join iPod? We've got some really cool shit. I've got this other project I really want to do but we need some time before we can convince Steve to do it, and I think you'd be great for it.'"

Grignon is boisterous and hardworking. He's also got a mouth like a Silicon Valley sailor. "So I left," Grignon says, "to work on this mystery thing. So we just kind of spun our wheels on some wireless speakers and shit like that, but then the project started to materialize. Of course what Fadell was talking about was the phone." Fadell knew Jobs was beginning to come around to the idea, and he wanted to be prepared. "We had this idea: Wouldn't it be great to put Wi-Fi in an iPod?" Grignon says. Throughout 2004, Fadell, Grignon, and the rest of the team worked on a number of early efforts to fuse iPod and internet communicator.

"That was one of the very first prototypes I showed Steve. We gutted an iPod, we had hardware add in a Wi-Fi part, so it was a big plastic piece of junk, and we modified the software." There were click-wheel iPods that could clumsily surf the web as early as 2004. "You would click the wheel, you would scroll the web page, and if there was a link on the page, it would highlight it, and you could click on it and you could jump in," Grignon says. "That was the very first time where we started experimenting with radios in the form factor."

It was also the first time Steve Jobs had seen the internet running on an iPod. "And he was like, 'This is bullshit.' He called it right away. . . . 'I don't want this. I know it works, I got it, great, thanks, but this is a shitty experience,'" Grignon says.

Meanwhile, Grignon says, "The exec team was trying to convince Steve that building a phone was a great idea for Apple. He didn't really see the path to success."

One of those trying to do the convincing was Mike Bell. A veteran of Apple, where he'd worked for fifteen years, and of Motorola's wireless division, Bell was positive that computers, music players, and cell phones were heading toward an inevitable convergence point. For months, he lobbied Jobs to do a phone, as did Steve Sakoman, a vice president who had worked on the ill-fated Newton.

"We were spending all this time putting iPod features in Motorola phones," Bell says. "That just seemed ass-backwards to me. If we just took the iPod user experience and some of the other stuff we were working on, we could own the market." It was getting harder to argue with that logic. The latest batches of MP3 phones were looking increasingly like iPod competitors, and new alternatives for dealing with the carriers were emerging. Meanwhile, Bell had seen Jony Ive's latest iPod designs, and he had some iPhone-ready models.

On November 7, 2004, Bell sent Jobs a late-night email. "Steve, I know you don't want to do a phone," he wrote, "but here's why we should do it: Jony Ive has some really cool designs for future iPods that no one has seen. We ought to take one of those, put some Apple software around it, and make a phone out if ourselves instead of putting our stuff on other people's phones."

Jobs called him right away. They argued for hours, pushing back and forth. Bell detailed his convergence theory—no doubt mentioning the fact that the mobile phone market was exploding worldwide—and Jobs picked it apart. Finally, he relented.

"Okay, I think we should go do it," he said.

"So Steve and I and Jony and Sakoman had lunch three or four days later and kicked off the iPhone project."

ENRI Redux

At 2 Infinite Loop, the touchscreen-tablet project was still chugging along. Bas Ording, Imran Chaudhri, and company were still exploring the contours of a basic touch-focused user interface.

One day, Bas Ording got a call from Steve. He said, "We're gonna do a phone."

Jobs hadn't forgotten about the multitouch interaction demos and the Q79 tablet project, but a tangle of obstacles — not least of which was that it was too expensive — shut it down. ("You've got to give me something I can sell," he told Imran.) But with a smaller screen and scaled-down system, Q79 might work as a phone.

"It's gonna have a small screen, it's gonna be just a touchscreen, there's not gonna be any buttons, and everything has to work on that," Jobs told Ording. He asked the UI wiz to make a demo of scrolling through a virtual address book with multitouch. "I was super-excited," Ording says. "I thought, *Yeah, it seems kind of impossible, but it would be fun to just try it.*" He sat down, "moused off" a phone-size section of his Mac's screen, and used it to model the iPhone surface. Those years in the touchscreen wilderness were paying off.

"We already had some other demos, a web page, for example — it was just a picture you could scroll with momentum," Ording says. "That's sort of how it started." The famous effect where your screen bounces when you hit the top or bottom of a page was born because Ording couldn't tell when he'd hit the top of a page. "I thought my program wasn't running because I tried to scroll and nothing would happen," he says, and then he'd realize he was scrolling in the wrong direction. "So that's when I started to think, *How can I make it so you can see or feel that you're at the end?* Right? Instead of feeling dead, like it's not responding."

These small details, which we now take for granted, were the product of exhaustive tinkering, of proof-of-concept experimenting. Like inertial scrolling, the wonky-sounding but now-universal effect that makes it look like and feel like you're flipping through a Rolodex when you scroll down your contact list.

"I had to try all kinds of things and figure out some math," Ording says. "Not all of it was complicated, but you have to get to the right combinations, and that's the tricky thing. " Eventually, Ording got it to feel natural.

"He called me back a few weeks later, and he had inertial scrolling working," Jobs said. "And when I saw the rubber band, inertial scrolling, and a few of the other things, I thought, 'My God, we can build a phone out of this.'"

★ ★ ★

Scott Forstall walked into Greg Christie's office near the end of 2004 and gave him the news too: Jobs wanted to do a phone. He'd been waiting about a decade to hear those words.

Christie is intense and brusque; his stocky build and sharp eyes feel loaded with kinetic energy. He joined Apple in the 1990s, when the company was in a downward spiral, just to work on the Newton—then one of the most promising mobile devices on the market. Then, he'd even tried to push Apple to do a Newton phone. "I'm sure I proposed it a dozen times," Christie says. "The internet was popping too—this is going to be a big deal: mobile, internet, phone."

Now, his Human Interface team—his knobs-and-dials crew—was about to embark on its most radical challenge yet. Its members gathered on the second floor of 2 Infinite Loop, right above the old user-testing lab, and set to work expanding the features, functionality, and look of the old ENRI tablet project. The handful of designers and engineers set up shop in a drab office replete with stained carpet, old furniture, a leaky bathroom next door, and little on the walls but a whiteboard and, for some reason, a poster of a chicken.

Jobs liked the room because it was secure, windowless, tucked away from straying eyes. The CEO was already imbuing the nascent iPhone project with top-to-bottom secrecy. "You know, the cleaning crews weren't allowed in here because there were these sliding whiteboards along the wall," Christie says. The team would sketch ideas on them, and the good ones stayed put. "We wouldn't erase them. They became part of the design conversation."

That conversation was about how to blend a touch-based UI with smartphone features.

Fortunately, they'd had a head start. There were the ENRI crew's multitouch demos, of course. But Imran Chaudhri had also led the design for Dashboard, which was full of widgets—weather, stocks, calculator, notes, calendar—that would be ideal for the phone. "The early idea for the phone was all about having these widgets in your pocket," Chaudhri says. So they ported them over.

The original design for many of those icons was actually created in a single night, back during the development of Dashboard. "It was one of those fucking crazy Steve deadlines," Imran says, "where he wanted to see a demo of everything." So he and Freddy Anzures, a recent hire to the HI team, spent a long night coming up with the rectilinear design concepts for the widgets—which would, years later, become the designs for the iPhone icons. "It's funny, the look of smartphone icons for a decade to come was hashed out in a few hours."

And they had to establish the fundamentals; for instance, What should it look like when you fire up your phone? A grid of apps seems like the obvious way to organize a smartphone's functions today— now that it's like water, as Chaudhri says—but it wasn't a foregone conclusion. "We tried some other stuff," Ording says. "Like, maybe it's a list of icons with the names after them." But what came to be called Springboard emerged early on as the standard. "They were little Chiclets, basically," Ording says. "Now, that's Imran too, that was a great idea, and it looked really nice."

Chaudhri had the Industrial Design team make a few wooden iPhone-like mock-ups so they could figure out the optimal size of the icons for a finger's touch.

The multitouch demos were promising, and the style was coming together. But what the team lacked was cohesion—a united idea of what a touch-based phone would be.

"It was really just sketches," Christie says. "Little fragments of ideas, like tapas. A little bit of this, a little of that. Could be part of Address Book, a slice of Safari." Tapas wouldn't sate Jobs, obviously; he wanted a full course. So he grew increasingly frustrated with the presentations.

"In January, in the New Year, he blows a gasket and tells us we're not getting it," Christie says. The fragments might have been impressive, but there was no narrative drawing the disparate parts together; it was a jumble of half-apps and ideas. There was no story.

"It was as if you delivered a story to your editor and it was a couple of sentences from the introductory paragraph, a few from the body, and then something from the middle of the conclusion—but not the concluding statement."

It simply wasn't enough. "Steve gave us an ultimatum," Christie recalls. "He said, You have two weeks. It was February of 2005, and we kicked off this two-week death march."

So Christie gathered the HI team to make the case that they should all march with him.

"Doing a phone is what I always wanted to do," he said. "I think the rest of you want to do this also. But we've got two weeks for one last chance to do this. And I really want to do it."

He wasn't kidding. For a decade, Christie had believed mobile computing was destined to converge with cell phones. This was his opportunity not only to prove he was right, but to drive the spark.

The small team was on board: Bas, Imran, Christie, three other designers—Stephen LeMay, Marcel van Os, and Freddy Anzures—and a project manager, Patrick Coffman. They worked around the clock to tie those fragments into a full-fledged narrative.

"We basically went to the mattresses," Christie says. Each designer was given a fragment to realize—an app to flesh out—and the team spent two sleepless weeks perfecting the shape and feel of an inchoate iPhone. And at the end of the death march, something resembling the one device emerged from the exhausted fog of the HI floor.

"I have no doubt that if I could resurrect that demo and show it to you now, you would have no problem recognizing it as an iPhone," Christie says. There was a home button—still software-based at this point—scrolling, and the multitouch media manipulations.

"We showed Steve the outline of the whole story. Showed him the home screen, showed him how a call comes in, how to go to your Address Book, and 'this is what Safari looks like,' and it was a little click through. It wasn't just some clever quotes, it told a story."

And Steve Jobs did love a good story.

"It was a smashing success," Christie says. "He wanted to go through it a second time. Anyone who saw it thought it was great. It *was* great."

It meant that the project was immediately deemed top secret. After the February demo, badge readers were installed on either end of the Human Interface group's hallway, on the second floor of 2 Infinite Loop. "It was lockdown," Christie says. "That's what you say when there's a prison riot, right? That was the phrase. Yeah, we're on lockdown."

It also meant they had a lot more work to do. If the ENRI meetings were prologue, the tablet prototyping the beginning, then this was the second act of the iPhone, and there was much left to be written. But now that Jobs was invested in the narrative, he wanted to show it off, in high style, to the rest of the company. "We had this 'big demo' — that's what we called it," Ording says. Steve wanted to show the iPhone prototype at the Top 100 meeting inside Apple. "They have this meeting every once in a while with all the important people, saying what the direction of the company is," Ording tells me. Jobs would invite the people he considered his top one hundred employees to a secret retreat, where they'd present and discuss upcoming products and strategies. For rising Applers, it was a make-or-break career opportunity. For Jobs, the presentations had to be as carefully calibrated as public-facing product launches.

"From then until May, it was another brutal haul, to, well . . . come up with connecting paragraphs," Christie says. "Okay, what are the apps we're going to have? What should a calendar in your hand look like. Email? Every step on this journey was just making it more and more concrete and more real. Playing songs out of your iTunes. Media playback. iPhone software started as a design project in my hallway with my team." Christie hacked the latest model of the iPod so the designers could get a feel for what the applications might look like on a device. The demo began to take shape. "You could tap on the mail app and see how that kind of works, and the web browser," Ording says. "It wasn't fully working, but enough that you could get the idea."

Christie uses one word to describe how the team toiled around the clock, you might have noticed, above all others. It was "brutal, grueling work. I put people in hotel rooms because I didn't want them driving home. People crashed at my house," he says, but "it was exhilarating at the same time."

Steve Jobs had been blown away by the results. And soon, so was everyone else. The presentation at Top 100 was another smash success.

The Bod of an iPod

When Fadell heard that a phone project was taking shape, he grabbed his own skunkworks prototype design of the iPod phone before he headed into an executive meeting.

"There was a meeting where they were talking about the formation of the phone project on the team," Grignon says. "Tony had [it] in his back pocket, a team already working on the hardware and the schematics, all the design for it. And once they got the approval for it from Steve, Tony was like, 'Oh, hold on, as a matter of fact' — *whoo-chaa!* Like he whipped it out, 'Here's this prototype that we've been thinking about,' and it was basically a fully baked design."

On paper, the logic looks impeccable: The iPod was Apple's most successful product, phones were going to eat the iPod's lunch, so why not an iPod phone? "Take the best of the iPod and put a phone in it," Fadell says. "So you could do mobile communications and have your music with you, and we didn't lose all the brand awareness we'd built into the iPod, the half a billion dollars we were spending getting that known around the world." It was that simple.

Remember that while it was becoming clear inside Apple that

they were going to pursue a phone, it wasn't clear at all what that phone should look or feel like. Or how it would work, on just about every level.

"Early 2005, around that time frame, Tony started saying there's talk about them doing a phone," says David Tupman, who was in charge of iPod hardware at the time. "And I said, 'I really want to do a phone. I'd like to lead that.' He said, 'No.'" Tupman laughs. "'You can't do that.' But they did a bunch of interviews, and I guess they couldn't find anybody, so I was like, 'Hello, I'm still here!' Tony was like, 'Okay, you're it.'"

The iPod team wasn't privy to what had been unfolding in the HI group.

"We were gonna build what everyone thought we should build at the time: Let's bolt a phone onto an iPod," says Andy Grignon. And that's exactly what they started to do.

What's It Going to Be?

Richard Williamson found himself in Steve Jobs's office. He'd gone in to discuss precisely the kind of thing that nobody wanted to discuss with Steve Jobs—leaving Apple.

For years, he had been in charge of the team that developed the framework that powered Safari, called WebKit. Here's a fun fact about WebKit: Unlike most products developed and deployed by Apple, it's open-source. Here's another: Until 2013, Google's own Chrome browser was powered by WebKit too. It's big-deal software, in other words. And Williamson was, as *Forbes* put it, "what's commonly referred to as a '@#$ rock star' in Silicon Valley." But he was getting burned out on upgrading the same platform.

"We had gone through three or four versions of WebKit, and I was thinking of moving to Google," he says. "That's when Steve invited me."

And Steve wasn't happy.

* * *

When you think "successful computer engineer," the stock photo that springs to mind is pretty much what Williamson looks like— bespectacled, unrepentantly geekish, brainy, wearing a button-down shirt. We met for an interview at a Palo Alto sushi joint that eschewed waiters in favor of automated service via table-mounted iPads. Seemed fitting.

Williamson is soft-spoken, with a light British accent. He seems affable but shy—there's a slightly anxious undercurrent to his speech—and unmistakably sharp. He's apt to rattle off ideas pulled from a deep knowledge of code, industry acumen, and the philosophy of technology, sometimes in the same breath. He was born in the United Kingdom in 1966, and his family moved to Phoenix when he was still a kid. "I started programming when I was about eleven," he says. "My dad worked for Honeywell, which at the time, they made mainframe computers.... Back then, pretty much the only way to access a mainframe was through a teletype terminal, and my dad had one at home. And it was one of these big old teletype printers," he recalls. "And you would dial up the mainframe by an acoustic modem—these old things where you stuck the telephone into a socket, through the receiver and a transmitter, and it would make these squirreling noises."

He was hooked. "I spent hours and hours programming—you know, I was totally a nerd, a geek." He wrote his own text-based adventure game. But his all-consuming computer habit was creating some problems: "I spent so much time on it that I ran out of paper." And eleven-year-olds can't afford to buy reams of computer paper. "I actually hacked the spooling system on the mainframe, which would spool large printouts, and you could get this spooled print job mailed to locations," he says. "So I would print out reams and reams of paper and mail them to my house so that I could keep on using the printer to access the mainframe."

At college, his skill set led a professor to ask him for help setting up the first computer science courses. "That's where I learned how to make a computer do pretty much whatever I wanted it to do," he says. A friend convinced him to start a company writing software for the Commodore Amiga, an early PC. "We wrote a program called Marauder, which was a program to make archival backups of copy-protected disks." He laughs. "That's kind of the diplomatic way of describing the program." Basically, they created a tool that allowed users to pirate software. "So we had a little bit of a recurring revenue stream," he says slyly.

In 1985, Steve Jobs's post-Apple company, NeXT, was still a small operation, and hungry for good engineers. There, Williamson met with two NeXT officers and one Steve Jobs. He showed them the work that he'd done on the Amiga, and they hired him on the spot. The young programmer would go on to spend the next quarter of a century in Jobs's—and the NeXT team's—orbit, working on the software that would become integral to the iPhone.

★ ★ ★

"Don't leave," Jobs said, according to Williamson. "We've got a new project I think you might be interested in."

So Williamson asked to see it. "At this point, there was nobody on the project from a software perspective, it was all just kind of an idea in Steve's mind." It didn't seem like a convincing reason for Williamson to pass up an enticing new offer. "Google was interested in giving me some very interesting work too, so it was a very pivotal moment," he says.

"So I said, 'Well, the screen isn't there, the display tech is kind of not really there.' But Steve convinced me it was. That the path would be there." Williamson pauses for a second. "It's all true about Steve," Williamson says with a quick smile. "I was with him since NeXT, and I've fallen under his glare many times."

What would it be, then? Of course, Williamson would stay. "So I became an advocate at that point of building a device to browse the web."

Which Phone

"Steve wanted to do a phone, and he wanted to do it as fast as he could," Williamson says. But which phone?

There were two options: (a) take the beloved, widely recognizable iPod and hack it to double as a phone (that was the easier path technologically, and Jobs wasn't envisioning the iPhone as a mobile computing device but as a souped-up phone), or (b) transmogrify a Mac into a tiny touch-tablet that made calls (which was an exciting idea but frayed with futuristic abstraction).

"After the big demo," Ording says, "the engineers started to look into, What would it take to actually make this real? On the hardware side but also the software side," Ording says. To say the engineers who first examined it were skeptical about its near-term viability would be an understatement. "They went, 'Oh my God, this is — we don't know, this is going to be a lot of work. We don't even know how much work.'"

There was so much that needed to be done to translate the multitouch Mac mass into a product, and one with so many new, unproven technologies, that it was difficult even to put forward a roadmap, to conceive of all of its pieces coming together.

For Those About to Rokr

Development had continued on the Rokr throughout 2005. "We all thought the Rokr was a joke," Williamson says. The famously hands-on CEO didn't see the finished Rokr until early September 2005, right before he was supposed to announce it to the world. And he was aghast. "He was like, 'What else can we do, how can we fix it?' He knew it was subpar but he didn't know how bad it was going to be. When it finally got there, he didn't even want to show it onstage because he was so embarrassed by it," Fadell says.

During the demonstration, Jobs held the phone like an unwashed sock. At one point the Rokr failed to switch from making calls to playing music, leaving Jobs visibly agitated. So, at about the same moment that Jobs was announcing "the world's first mobile phone with iTunes" to the media, he was resolving to make it obsolete.

"When he got offstage he was just like, 'Ugh,' really upset," Fadell says. The Rokr was such a disaster that it landed on the cover of *Wired* with the headline "You Call This the Phone of the Future?" and it was soon being returned at a rate six times higher than the industry average. Its sheer shittiness took Jobs by surprise—and his anger helped motivate him to squeeze the trigger harder on an Apple-built phone. "It wasn't when it failed. It was right after it launched," Fadell says. "This is not gonna fly. I'm sick and tired of dealing with bozo handset guys," Jobs told Fadell after the demo.

"That was the ultimate thing," Fadell says. "It was, 'Fuck this, we're going to make our own phone.'"

* * *

"Steve called a big meeting in the boardroom," Ording says. "Everyone was there, Phil Schiller and Jony Ive and whoever." He said, "Listen. We're going to change plans.... We're going to do this iPod-based thing, make that into a phone because that's a much more doable project. More predictable." That was Fadell's project. The touchscreen effort wasn't abandoned, but while the engineers worked on whipping it into shape, Jobs directed Ording, Chaudhri, and members of the UI team to design an interface for an iPod phone, a way to dial numbers, select contacts, and browse the web using that device's tried-and-true click wheel.

There were now two competing projects vying to become the iPhone—a "bake-off," as some engineers put it. The two phone projects were split into tracks, code-named P1 and P2, respectively. Both were top secret. P1 was the iPod phone. P2 was the still-experimental hybrid of multitouch technology and Mac software.

If there's a ground zero for the political strife that would later come to engulf the project, it's likely here, in the decision to split the two teams—Fadell's iPod division, which was still charged with updating that product line in addition to prototyping the iPod phone, and Scott Forstall's Mac OS software vets—and drive them to compete. (The Human Interface designers, meanwhile, worked on both P1 and P2.)

Eventually, the executives overseeing the most important elements of the iPhone—software, hardware, and industrial design—would barely be able to tolerate sitting in the same room together. One would quit, others would be fired, and one would emerge solidly—and perhaps solely—as the new face of Apple's genius in the post-Jobs era. Meanwhile, the designers, engineers, and coders would work tirelessly, below the political fray, to turn the Ps into working devices in any way possible.

The Purple People Leader

Every top secret project worth its salt in intrigue has a code name. The iPhone's was Purple.

"One of the buildings we have up in Cupertino, we locked it down," said Scott Forstall, who had managed Mac OS X software and who would come to run the entire iPhone software program. "We started with one floor"—where Greg Christie's Human Interface team worked—"We locked the entire floor down. We put doors with badge readers, there were cameras, I think, to get to some of our labs, you had to badge in four times to get there." He called it the Purple Dorm because, "much like a dorm, people were there all the time."

They "put up a sign that said 'Fight Club' because the first rule of Fight Club in the movie is that you don't talk about Fight Club, and the first rule about the Purple Project is you do not talk about that outside of those doors," Forstall said.

Why Purple? Few seem to recall. One theory is it was named after a purple kangaroo toy that Scott Herz—one of the first engineers to come to work on the iPhone—had as a mascot for Radar, the system that Apple engineers used to keep track of software bugs and glitches throughout the company. "All the bugs are tracked inside of Radar at Apple, and a lot of people have access to Radar," says Richard Williamson. "So if you're a curious engineer, you can go spelunking around the bug-tracking system and find out what people are working on. And if you're working on a secret project, you have to think about how to cover your tracks there."

Scott Forstall, born in 1969, had been downloading Apple into his brain his entire life. By junior high, his precocious math and science skills landed him in an advanced-placement course with access to an Apple IIe computer. He learned to code, and to code well. Forstall didn't fit the classic computer-geek mold, though. He was a debate-team champ and a performer in high-school musicals; he played the lead in *Sweeney Todd,* that hammy demon barber. Forstall graduated from Stanford in 1992 with a master's in computer science and landed a job at NeXT.

After releasing an overpriced computer aimed at the higher-education market, NeXT flailed as a hardware company, but it survived by licensing its powerful NeXTSTEP operating system. In 1996, Apple bought NeXT and brought Jobs back into the fold, and the decision was made to use NeXTSTEP to overhaul the Mac's aging operating system. It became the foundation on which Macs—and iPhones—still run today. At Jobs-led Apple, Forstall rose through the ranks. He mimicked his idol's management style and distinctive taste. *BusinessWeek* called him "the Sorcerer's Apprentice."

One of his former colleagues praised him as a smart, savvy leader but said he went overboard on the Jobs-worship: "He was generally great, but sometimes it was like, just be yourself." Forstall emerged as the leader of the effort to adapt Mac software to a touchscreen phone. Though some found his ego and naked ambition distasteful—he was "very much in need of adulation," according to one peer, and called "a starfucker" by another—few dispute the caliber of his intellect and work ethic. "I don't know what other people have said about Scott," Henri Lamiraux says, "but he was a pleasure to work with."

Forstall led many of the top engineers he'd worked with since his NeXT days—Henri Lamiraux and Richard Williamson among them—into the P2 project. Williamson jokingly called the crew "the NeXT mafia." True to the name, they would at times behave in a manner befitting a close-knit, secretive (and highly efficient) organization.

P1 Thing After Another

Tony Fadell was Forstall's chief competition.

"From a politics perspective, Tony wanted to own the entire experience," Grignon says. "The software, the hardware ... once people started to see the importance of this project to Apple, everyone wanted to get their fingers in it. And that's when the epic fight between Fadell and Forstall began."

Having worked with Forstall on Dashboard, Grignon was in a unique position to interface with both groups. "From our perspective, Forstall and his crew, we always viewed them as the underdogs. Like they were trying to wedge their way in," Grignon says. "We had complete confidence that our stack was going to happen because this is Tony's project, and Tony's responsible for millions upon millions of iPod sales."

So, the pod team worked to produce a new pod-phone cut from the mold of Apple's ubiquitous music player. Their idea was to produce an iPod that would have two distinct modes: the music player and a phone. "We prototyped a new way," Grignon says of the early device. "It was this interesting material ... it still had this touch sensitive click wheel, right, and the Play/Pause/Next/Previous buttons in blue backlighting. And when you put it into phone mode through the UI, all that light kind of faded out and faded back in as orange. Like, zero to nine in the click wheel in an old rotary phone, you know, ABCDEGF around the edges." When the device was in music-playing mode, blue backlighting would show iPod controls around the touch wheel. The screen would still be filled with iPod-style text and lists, and if you toggled it to phone mode, it'd glow orange and display numbers like the dial of a rotary phone.

"We put a radio inside, effectively an iPod Mini with a speaker and headphones, still using the touch-wheel interface," Tupman says.

"And when you texted, it dialed—and it worked!" Grignon says. "So we built a couple hundred of them."

The problem was that they were difficult to use as phones. "After we made the first iteration of the software, it was clear that this was going nowhere," Fadell says. "Because of the wheel interface. It was never gonna work because you don't want a rotary dial on the phone."

The design team tried mightily to hack together a solution.

"I came up with some ideas for the predictive typing," Bas Ording says. There would be an alphabet laid out at the bottom of the screen, and users would use the wheel to select letters. "And then you can just, like, *click-click-click-click* — 'Hello, how are you.' So I just built an actual thing that can learn as you type—it would build up a database of words that follow each other." But the process was still too tedious.

"It was just obvious that we were overloading the click wheel with too much," Grignon says. "And texting and phone numbers—it was a fucking mess."

"We tried everything," Fadell says. "And nothing came out to make it work. Steve kept pushing and pushing, and we were like, 'Steve.' He's pushing the rock up a hill. Let's put it this way: I think he knew, I could tell in his eyes that he knew; he just wanted it to work," he says. "He just kept beating this dead horse."

"C'mon, there's gotta be a way," Jobs would tell Fadell. "He didn't just want to give up. So he pushed until there was nothing there," Fadell says.

They even filed for a patent for the ill-fated device, and in the bowels of Cupertino, there were offices and labs littered with dozens of working iPod phones. "We actually made phone calls," Grignon says.

The first calls from an Apple phone were not, it turns out, made on the sleek touchscreen interface of the future but on a steampunk rotary dial. "We came very close," Ording says. "It was, like, we could have finished it and made a product out of it....But then I guess Steve must have woken up one day like, 'This is not as exciting as the touch stuff.'"

"For us on the hardware team, it was great experience," David Tupman says. "We got to build RF radio boards, it forced us to select suppliers, it pushed us to get everything in place." In fact, elements of the iPod phone wound up migrating into the final iPhone; it was like a version 0.1, Tupman says. For instance: "The radio system that was in that iPod phone was the one that shipped in the actual iPhone."

Hands Off

The first time Fadell saw P2's touch-tablet rig in action, he was impressed—and perplexed. "Steve pulled me in a room when everything was failing on the iPod phone and said, 'Come and look at this.'" Jobs showed him the ENRI team's multitouch prototype. "They had been getting, in the background, the touch Mac going. But it wasn't a touch Mac; literally, it was a room with a Ping-Pong table, a projector, and this thing that was a big touchscreen," Fadell says.

"This is what I want to put on the phone," Jobs said.

"Steve, sure," Fadell replied. "It's not even close to production. It's a prototype, and it's not a prototype at scale—it's a prototype table. It's a research project. It was like eight percent there," Fadell says.

David Tupman was more optimistic. "I was like, 'Oh, wow, yeah, we *have* to find out a way to make this work.'" He was convinced the engineering challenges could be solved. "I said, 'Let's just sit down and go through the numbers and let's work it out.'"

The iPod phone was losing support. The executives debated which project to pursue, but Phil Schiller, Apple's head of marketing, had an answer: Neither. He wanted a keyboard with hard buttons.

The BlackBerry was arguably the first hit smartphone. It had an email client and a tiny hard keyboard. After everyone else, including Fadell, started to agree that multitouch was the way forward, Schiller became the lone holdout.

He "just sat there with his sword out every time, going, 'No, we've got to have a hard keyboard. No. Hard keyboard.' And he wouldn't listen to reason as all of us were like, 'No, this works now, Phil.' And he'd say, 'You gotta have a hard keyboard!'" Fadell says.

Schiller didn't have the same technological acumen as many of the other execs. "Phil is not a technology guy," Brett Bilbrey, the former head of Apple's Advance Technology Group, says. "There were days when you had to explain things to him like a grade-school kid." Jobs liked him, Bilbrey thinks, because he "looked at technology like middle America does, like Grandma and Grandpa did."

When the rest of the team had decided to move on multitouch and a virtual keyboard, Schiller put his foot down. "There was this one spectacular meeting where we were finally going in a direction," Fadell says, "and he erupted."

"We're making the wrong decision!" Schiller shouted.

"Steve looked at him and goes, 'I'm sick and tired of this stuff. Can we get off of this?' And he threw him out of the meeting," Fadell recalls. Later, he says, "Steve and he had it out in the hallway. He was told, like, Get on the program or get the fuck out. And he ultimately caved."

That cleared it up: the phone would be based on a touchscreen. "We all know this is the one we want to do," Jobs said in a meeting, pointing to the touchscreen. "So let's make it work."

Round Two

"There was a whole religious war over the phone" between the iPod team and the Mac OS crew, one former Apple executive told me. When the iPod wheel was ruled out and the touch ruled in, the new question was how to build the phone's operating system. This was a critical juncture—it would determine whether the iPhone would be positioned as an accessory or as a mobile computer.

"Tony and his team were arguing we should evolve the operating system and take it in the direction of the iPod, which was very rudimentary," Richard Williamson says. "And myself and Henri and Scott Forstall, we were all arguing we should take OS X"—Apple's main operating system, which ran on its desktops and laptops—"and shrink it down."

"There were some epic battles, philosophical battles about trying to decide what to do," Williamson says.

The NeXT mafia saw an opportunity to create a true mobile computing device and wanted to squeeze the Mac's operating system onto the phone, complete with versions of Mac apps. They knew the operating system inside and out—it was based on code they'd worked with for over a decade. "We knew for sure that there was enough horsepower to run a modern operating system," Williamson says, and they believed they could use a compact ARM processor—Sophie Wilson's low-power chip architecture—to create a stripped-down computer on a phone.

The iPod team thought that was too ambitious and that the phone should run a version of Linux, the open-source system popular with developers and open-source advocates, which already ran on low-power ARM chips. "Now we've built this phone," says Andy Grignon, "but we have this big argument about what was the operating system it should be built on. 'Cause we were initially making it iPod-based, right? And nobody cares what the operating system in an iPod is. It's an appliance, an accessory. We were viewing the phone in that same camp."

Remember, even after the iPhone's launch, Steve Jobs would describe it as "more like an iPod" than a computer. But those who'd been in the trenches experimenting with the touch interface were excited about the possibilities it presented for personal computing and for evolving the human-machine interface.

"There was definitely discussion: This is just an iPod with a phone. And we said, no, it's OS X with a phone," Henri Lamiraux says. "That's what created a lot of conflict with the iPod team, because they thought they were the team that knew about all the software on small devices. And we were like, no, okay, it's just a computer."

"At this point we didn't care about the phone at all," Williamson says. "The phone's largely irrelevant. It's basically a modem. But it was 'What is the operating system going to be like, what is the inter-action paradigm going to be like?'" In that comment, you can read the roots of the philosophical clash: The software engineers saw P2 not as a chance to build a phone, but as an opportunity to use a phone-shaped device as a Trojan horse for a much more complex kind of mobile computer.

The Incredible Shrinking Operating System

When the two systems squared off early on, the mobile-computing approach didn't fare so well.

"Uh, just the load time was laughable," Andy Grignon says. Grignon's Linux option was fast and simple. "It's just kind of *prrrrrt* and it's up." When the Mac team first got their system compiling, "it was like six rows of hashtags, *dink-dink-dink-dink-dink*, and then it just sat there and it would shit the bed for a little bit, and then it would finally come back up and you'd be like, Are you even kidding me? And this is supposed to be for a device that just turns on? Like, for real?"

"At that point it was up to us to prove" that a variant of OS X could work on the device, Williamson says. The mafia got to work, and the competition heightened. "We wanted our vision for this phone that Apple was going to release to become a reality," Nitin Ganatra says. "We didn't want to let the iPod team have an iPod-ish version of the phone come out before."

One of the first orders of business was to demonstrate that the scrolling that had wowed Jobs would work with the stripped-down operating system. Williamson linked up with Ording and hashed it out. "It worked and looked amazingly real. When you touched the screen, it would track your finger perfectly, you would pull down, it would pull down."

That, Williamson says, put the nail in the Linux pod's coffin. "Once we had OS X ported and these basic scrolling interactions nailed, the decision was made: We're not going to go with the iPod stack, we're going to go with OS X."

The software for the iPhone would be built by Scott Forstall's NeXT mafia; the hardware would go to Fadell's group. The iPhone would boast a touchscreen and pack the power of a mobile computer. That is, if they could get the thing to work.

Fadell looked at the multitouch contraption again. "I didn't say, 'Sure,' and I didn't say, 'No.' I said, 'Okay, we've got a lot of things to work on,'" he says. "We had to create a whole basically separate company just to build the prototype."

★ ★ ★

Long after its launch, the iPhone would not only require the creation of such "separate companies" inside Apple, it would lead to the absorption of entirely new ones outside it. It would lead to new breakthroughs, new ideas, new hurdles. These next chapters engage with the world reorganized by the iPhone — from the ascent of Siri and the Secure Enclave, to the making, marketing, and trashing of the one device.

CHAPTER 10

Hey, Siri

Where was the first artificially intelligent assistant born?

Of all the places I might have expected to have a deep conversation with one of the founders of Siri about the evolving state of artificial intelligence, a cruise ship circling Papua New Guinea was not a frontrunner. But here we are, in my cabin on the *National Geographic Orion*, the buzz of its engine providing a suitably artificial thrum to the backdrop of our talk, a tropical green-blue expanse out the window.

"I'm going to have to watch what I say on the record here," says Tom Gruber with a short smile and a nod toward my recorder. That's because Gruber is head of advanced development for Siri at Apple. We're both aboard Mission Blue, a seafaring expedition organized by TED, the pop-lecture organization, and Sylvia Earle, the oceanographer, to raise awareness of marine-conservation issues. By night, there are TED Talks. By day, there's snorkeling.

Gruber's easy to spot—he's the goateed mad scientist flying the drone. He looks like he's constantly scanning the room for intel. He talks softly but at a whirring clip, often cutting one rapid-fire thought short to begin another. "I'm interested in the human interface," he says. "That's really where I put my effort. The AI is there to serve the human interface, in my book."

Right now, he's talking about Siri, Apple's artificially intelligent personal assistant, who probably needs little introduction.

Siri is maybe the most famous AI since HAL 9000, and "Hey, Siri" is probably the best-known AI-human interaction since "I'm sorry, Dave, I'm afraid I can't do that." One of those AIs, of course, assists us with everyday routines—Siri answered one billion user requests per week in 2015, and two billion in 2016—and the other embodies our deepest fears about machine sentience gone awry.

Yet if you ask Siri where she—sorry, *it,* but more on that in a second—comes from, the reply is the same: "I, Siri, was designed by Apple in California." But that isn't the full story.

Siri is really a constellation of features—speech-recognition software, a natural-language user interface, *and* an artificially intelligent personal assistant. When you ask Siri a question, here's what happens: Your voice is digitized and transmitted to an Apple server in the Cloud while a local voice recognizer scans it right on your iPhone. Speech-recognition software translates your speech into text. Natural-language processing parses it. Siri consults what tech writer Steven Levy calls the iBrain—around 200 megabytes of data about your preferences, the way you speak, and other details. If your question can be answered by the phone itself ("Would you set my alarm for eight a.m.?"), the Cloud request is canceled. If Siri needs to pull data from the web ("Is it going to rain tomorrow?"), to the Cloud it goes, and the request is analyzed by another array of models and tools.

Before Siri was a core functionality of the iPhone, it was an app on the App Store launched by a well-funded Silicon Valley start-up. Before that, it was a research project at Stanford backed by the Defense Department with the aim of creating an artificially intelligent assistant. Before that, it was an idea that had bounced around the tech industry, pop culture, and the halls of academia for decades; Apple itself had an early concept of a voice-interfacing AI in the 1980s.

Before *that* there was the Hearsay II, a proto-Siri speech-recognition system. And Gruber says it was the prime inspiration for Siri.

* * *

Dabbala Rajagopal "Raj" Reddy was born in 1937 in a village of five hundred people south of Madras, India. Around then, the region was hit with a seven-year drought and subsequent famine. Reddy learned to write, he says, by carving figures in the sand. He had difficulty with language, switching from his local dialect to English-only classes when he went to college, where professors spoke with Irish, Scottish, and Italian accents. Reddy headed to the College of Engineering at the University of Madras and afterward landed an internship in Australia. It was then, in 1959, that he first became acquainted with the concept of a computer.

He got a master's at the University of New South Wales, worked at IBM for three years, then moved to Stanford, where he'd eventually obtain his doctorate. He was drawn to the dawning study of artificial intelligence, and when his professor asked him to pick a topic to tackle, he gravitated to one in particular: speech recognition.

"I chose that particular one because I was both interested in languages, having come from India at that time, and in having had to learn three or four languages," he said in a 1991 interview for the Charles Babbage Institute. "Speech is something that is ubiquitous to humankind. . . . What I didn't know was that it was going to be a lifetime problem. I thought it was a class project."

Over the next few years, he tried to build a system for isolated word recognition—a computer that could understand the words humans spoke to it. The system he and his colleagues created in the late 1960s, he said, "was the largest that I knew of at that time—about 560 words or something—with a respectable performance of like about 92%." As with much of the advanced computer research around Stanford then, ARPA was doing the funding. It would mark a decades-long interest in the field of AI from the agency, which would fund multiple speech recognition projects in the 1970s. In 1969, Reddy moved to Carnegie Mellon and continued his work. There, with more ARPA funding, he launched the Hearsay project. It was, essentially, Siri, in its most embryonic form. "Ironically, it was a speech interface," Gruber says. "A Siri kind of thing. It was 1975, I think; it was something crazy."

Hearsay II could correctly interpret a thousand words of English a vast majority of the time.

* * *

"I just think the human mind is kind of the most interesting thing on the planet," Tom Gruber says. He went to Loyola University in New Orleans to study psychology before discovering he had a knack for computers, which were just emerging on the academic scene. When the school got a Moog synthesizer, he whipped up a computer interface for it. And he created a computer-aided instruction system that's still used at Loyola's psych department today. Then Gruber stumbled across a paper published by a group of scientists at Carnegie Mellon University—led by Raj Reddy.

What Gruber saw in the paper were the worm roots of AI—a speech-recognition system capable of symbolic reasoning. The beginnings of what would, decades later, become Siri. It's one thing to train a computer to recognize sounds and match them to data stored in a database. But Reddy's team was trying to figure out how the language could be represented within a computer so that the machine could do something useful with it. For that, it had to learn to be able to recognize and break down the different parts of a sentence.

Symbolic reasoning describes how the human mind uses symbols to represent numbers and logical relationships to solve problems both simple and complex.

Like: "'We have an appointment at two to have an interview,'" Gruber says, referring to the time we'd set aside for our talk. "That's a statement of fact that can be represented in knowledge-representation terms. It can't be represented as a database entry unless the entire database is nothing but instances of that fact." So you could, he's saying, set up a massive database of every possible date and time, teach the computer to recognize it, and play a matching game. "But that's not a knowledge representation. The knowledge representation is

'You human being, me human being. Meet at time and place. Maybe ostensibly for a purpose' — and that is the basis for intelligence."

Gruber graduated summa cum laude in 1981 and headed to grad school at the University of Massachusetts at Amherst, where he looked into ways that AI might benefit the speech-impaired. "My first project was a human-computer interface using artificial intelligence to help with what's called communication prosthesis," he says. The AI would analyze the words of people who suffered from speech-impeding afflictions like cerebral palsy, and predict what they were trying to say. "It is actually an ancestor of something that I call 'semantic autocomplete.'

"I used it later in Siri," Gruber says. "The same idea, just modernized."

★ ★ ★

The automated personal assistant is yet another one of our age-old ambitions and fantasies.

"He was fashioning tripods, twenty in all, to stand around the wall of his well-builded hall, and golden wheels had he set beneath the base of each that of themselves they might enter the gathering of the gods at his wish and again return to his house, a wonder to behold." That might be the earliest recorded imagining of an autonomous mechanical assistant, and it appears in Homer's *Iliad,* written in the eighth century B.C. Hephaestus, the Greek god of blacksmiths, has innovated a little fleet of golden-wheeled tripods that can shuttle to and from a god party upon command—Homeric robot servants.

Siri, too, is essentially a mechanical servant. As Bruce G. Buchanan, a founding member of the American Association of Artificial Intelligence, puts it, "The history of AI is a history of fantasies, possibilities, demonstrations, and promise." Before humans had anything approaching the technological know-how to create machines that emulated humans, they got busy imagining what might happen if they did.

Jewish myths of golems, summoned out of clay to serve as protectors and laborers, but which usually end up running amok, are centuries old. Mary Shelley's *Frankenstein* was an AI patched together with corpses and lightning. As far back as the third century B.C., a Lie Zie text describes an "artificer" presenting a king with a lifelike automaton, essentially a mechanical human mannequin that could sing and dance. The first time the word *robot* appeared was to describe the eponymous subjects of the playwright Karel Čapek's 1922 work *Rossum's Universal Robots*. Čapek's new word was derived from *robota*, which means "forced labor." Ever since, the word *robot* has been used to describe nominally intelligent machines that perform work for humans. From *The Jetsons'* robo-maid Rosie to the Star Wars droids, robots are, basically, mechanical assistants.

Primed by hundreds of years of fantasy and possibility, around the mid-twentieth century, once sufficient computing power was available, the scientific work investigating actual artificial intelligence began. With the resonant opening line "I propose to consider the question, 'Can machines think?'" in his 1950 paper "Computing Machinery and Intelligence," Alan Turing framed much of the debate to come. That work discusses his famous Imitation Game, now colloquially known as the Turing Test, which describes criteria for judging whether a machine may be considered sufficiently "intelligent." Claude Shannon, the communication theorist, published his seminal work on information theory, introducing the concept of the bit as well as a language through which humans might speak to computers. In 1956, Stanford's John McCarthy and his colleagues coined the term *artificial intelligence* for a new discipline, and we were off to the races.

Over the next decade, as the scientific investigation of AI began to draw interest from the public and as, simultaneously, computer terminals became a more ubiquitous machine-human interface, the two future threads—screen-based interfaces and AI—wound into one, and the servile human-shaped robots of yore became disembodied. In the first season of the original *Star Trek*, Captain Kirk speaks to a cube-shaped computer. And, of course, in *2001: A Space Odyssey*,

Hal 9000 is an omnipresent computer controlled—for a while—through voice commands.

"Now, Siri was more about traditional AI being the assistant," Gruber says. "The idea, the core idea of having an AI assistant, has been around forever. I used to show clips from the Knowledge Navigator video from Apple." The video he's referring to, which is legendary in certain tech circles, is a wonderfully odd bit of early design fiction from John Sculley–era Apple. It depicts a professor in an opulent, Ivy League–looking office consulting his tablet (even now, Gruber refers to it as a Dynabook, and Alan Kay was apparently a consultant on the Knowledge Navigator project) by speaking to it. His computer is represented by a bow-tie-wearing dandy who informs Professor Serious about his upcoming engagements and his colleagues' recent publications. "So that's a model of Siri; that was there in 1987."

★ ★ ★

Gruber's 1989 dissertation, which would be expanded into a book, was "The Acquisition of Strategic Knowledge," and it described training an AI assistant to acquire knowledge from human experts.

The period during which Gruber attended grad school was, he says, a "peak time when there were two symbolic approaches to AI. There was essentially pure logical representation and generic reasoning."

The logic-driven approach to AI included trying to teach a computer to reason using those symbolic building blocks, like the ones in an English sentence. The other approach was data-driven. That model says, "No, actually the problem is a representation of memory and reasoning is a small part," Gruber says. "So lawyers, for instance, aren't great lawyers because they have deep-thinking minds that solve riddles like Einstein's. What the lawyers are good at is knowing a lot of stuff. They have databases, and they can comb through them rapidly to find the correct matches, the correct solutions."

Gruber was in the logic camp, and the approach is "no longer fashionable. Today, people want no knowledge but lots of data and machine learning."

It's a tricky but substantive divide. When Gruber says *knowledge*, I think he means a firm, robust grasp on how the world works and how to reason. Today, researchers are less interested in developing AI's ability to reason and more intent on having them do more and more complex machine learning, which is not unlike automated data mining. You might have heard the term *deep learning*. Projects like Google's DeepMind neural network work essentially by hoovering up as much data as possible, then getting better and better at simulating desired outcomes. By processing immense amounts of data about, say, Van Gogh's paintings, a system like this can be instructed to create a Van Gogh painting—and it will spit out a painting that looks kinda-sorta like a Van Gogh. The difference between this data-driven approach and the logic-driven approach is that this computer doesn't *know* anything about Van Gogh or what an artist is. It is only imitating patterns—often very well—that it has seen before.

"The thing that's good for is perception," Gruber says. "The computer vision, computer speech, understanding, pattern recognition, and these things did not do well with knowledge representations. They did better with data- and signal-processing techniques. So that's what's happened. The machine learning has just gotten really good at making generalizations over training examples."

But there is, of course, a deficiency in that approach. "The machine-learned models, no one really has any idea of what the models know or what they mean; they just perform in a way that meets the objective function of a training set"—like producing the Van Gogh painting.

Scientists have a fairly good grasp on how human perception works—the mechanisms that allow us to hear and see—and those can be modeled pretty fluidly. They don't, of course, have that kind of understanding of how our brains work. There's no scientific consensus on how humans understand language, for instance. The databases can mimic how we hear and see, but not how we think. "So a lot of people think that's AI. But that's only the perception."

After Amherst, Gruber went to Stanford, where he invented hyper-mail. In 1994, he started his first company, Intraspect— "Essentially...a group mind for corporations." He spent the next decade or so bouncing between start-ups and research. And then he met Siri. Or, rather, he met what was about to become Siri. It had been a long time in the making.

* * *

Before we can get to Siri, we have get back to DARPA. The Defense Advanced Research Projects Agency (or ARPA before 1972) had funded a number of AI and speech-recognition projects in the 1960s, leading Raj Reddy and others to develop the field and inspiring the likes of Tom Gruber to join the discipline. In 2003, decades later, DARPA made an unexpected return to the AI game.

The agency gave the nonprofit research outfit SRI International around two hundred million dollars to organize five hundred top scientists in a concerted research effort to build a virtual AI. The project was dubbed Cognitive Assistant that Learns and Organizes, or CALO—an attempt to wring an acronym out of *calonis*, which is, somewhat ominously, Latin for "soldier's servant." By the 2000s, AI had fallen out of fashion as a research pursuit, so the large-scale effort took some in the field by surprise. "CALO was put together at a time when many people said AI was a waste of time," Paul Saffo, a technology forecaster at Stanford University, told the *Huffington Post*. "It had failed multiple times, skepticism was high and a lot of people thought it was a dumb idea."

One reason for the DoD's sudden interest in AI could have been the escalation of the Iraq War, which began in 2003—and indeed, some technology developed under CALO was deployed in Iraq as part of the army's Command Post of the Future software system. Regardless, AI went from semi-dormant to a field of major activity. CALO was, "by any measure, the largest AI program in history," said David Israel, one of its lead researchers. Some thirty universities sent their best AI researchers, and proponents of each of the major approaches to AI collaborated for the first time. "SRI had this project," Gruber says. "They were paid by the government two hundred million bucks to run this project that was creating . . . a sensing office assistant, like it'd help you with meetings and PowerPoint, stuff like that. They wanted to push the art of AI," he says.

After the project drew to a close in 2008, its chief architect, Adam Cheyer, and a key executive, Dag Kittlaus, decided to spin some of the fundamental elements of the research into a start-up.

"They came up with an architecture for How would you represent all the bits an assistant needs to know? Like, how do you recognize speech? How do you recognize human language? How do you understand service providers like Yelp or your calendar app and how do you combine the input with task intent?" Gruber says.

Cheyer and Kittlaus imagined their assistant as a navigator of a "do engine" that would supplant the search engine as the predominant way people got around online. Proto-Siri would not only be able to scan the web, but also send a car to come pick you up on command, for instance. Originally, however, it wasn't conceived as a voice interface, Gruber says.

"It was an assistant; it did language understanding. It didn't do speech recognition," he says. "It did some natural-language understanding as you type into it. But it was much more focused on things like scheduling and making a dossier on people when you meet them and things like that.

"It was a cool, cool project but it was made for people typing on computers."

Gruber was introduced to the project when it was still in its "early prototype brainstorming phase" and he met with the two co-founders. "And I said, that's a really good idea but this is a consumer play. . . . We need to make an interface for this," Gruber says. "My little tiny team inside of Siri created that conversational interface. So the whole way you see it now, the same paradigm everyone uses is these conversational threads, there's content in between." It's not just command and response, designed to be as efficient as possible. Siri *talks* to you. "There's dialogue to disambiguate. It's just this this notion of a verbal, to-and-fro assistant that came out of there."

The project had begun the year after the first iPhone launched, and as the Siri project took shape, it was clear that it would be aimed at smartphones. "The Siri play was mobile from the beginning," he says. "Let's do an assistant, let's make it mobile. And then let's add speech, when speech is ready....By the second year, the speech-recognition technology had gotten good enough we could license it."

Now Gruber and his colleagues had to think about how people might talk to an AI interface, something that had never really existed in the consumer market before. They would have to think about how to train people to know what a workable command was in Siri's eyes, or ears.

"We had to teach people what you could and couldn't say to it, which is still a problem, but we did a better job as a start-up than I think that we currently do," Gruber says. Siri would often be sluggish because it would take time to process commands and formulate a response. "The idea that Siri talks back to you and does snappy things and so on, that was an outgrowth of a problem of how do you deal with the fact that Siri can't know most things? So you fall back on either web search or on a thing that looks like Siri knows something when it doesn't." Siri is, basically, buying time. "Like, Siri pretends to talk to you as if it knows you as a person who doesn't really know that, but it's a good illusion." And that illusion becomes less necessary the more adapted to your voice Siri becomes.

They also had to think about how best to foster engagement, to get people interested in returning to Siri. "That's the thing—you want engagement," Gruber says. "So we use a relatively straightforward way of doing conversation, but we focus a lot on content, not just form.

"If you were given a thing to ask questions to, what would the top ten things be? And people ask, like, 'What's the meaning of life?' And 'Will you marry me?' And all that stuff. And very quickly, we saw what were going to be the top questions and then we wrote really good answers. I hired a brilliant guy to write dialogue." Gruber couldn't give me his name, because he still works at Apple, but all signs point to Harry Sadler, whose LinkedIn page lists him as Manager, Siri Conversational Interaction Design. Today, an entire team writes Siri's dialogue. And they spend a lot of time fine-tuning its tone.

"We designed the character not to be gender-specific, not to even be species-specific. To try to pretend like humans are this funny species," Gruber says. It "finds them humorous, and curious." Originally, Siri was more colorful — it dropped f-bombs, teased users more aggressively, had more of a bombastic personality. But it was an open question. What do we want our artificially intelligent personal assistant to sound like? Whom do we want to talk to every day, and how do we want to be talked to?

"I mean, it's a great problem, right?" he says. "You have this giant audience of people, and you just have to write snappy little things and they'll love it. Imagine you're writing a book, and you're developing a character. And you think about, what does this character do? Well, it's an assistant that doesn't really know human culture, is curious about it, but it does its very best, it's professional. You can insult it, but it's not going to take shit. But it's not going to fight you either... that's the thing it has to be, because Apple's not going to put out quotable offensive things, even comebacks, even though we can write them. So that was really a fine art, to write that stuff."

Whoever designed it, Gruber credits him with perfecting that character. "He eventually owned it — he created the dialogue tone. As the writer. He really understood that you need a personality."

Still, someone would need to give voice to that personality. That someone was Susan Bennett, a sixty-eight-year-old voice actor who lived in the Atlanta suburbs. In 2005, Bennett spent every day of July recording every utterable word and vowel for a company called Scan-Soft. It was arduous, tedious work. "There are some people that just can read hour upon hour upon hour, and it's not a problem. For me, I get extremely bored," Bennett said. It was also hard to retain the android-esque monotone for hours at a time. "That's one of the reasons why Siri might sometimes sound like she has a bit of an attitude." ScanSoft rebranded itself as Nuance, and the pre-Apple Siri bought its voice-recognition system—and Siri's voice—for the app. She had no idea she was about become the voice of AI—she didn't find out she was Siri until 2011, when someone emailed her. Apple won't confirm Bennett's involvement, though speech analysts have reported that it's her. "I had really ambivalent feelings," she said. "I was flattered to be chosen to basically be the voice of Apple in North America, but having been chosen without my knowledge was strange. Especially since my voice was on millions and millions of devices."

★ ★ ★

"Even the name was very carefully culturally tested," Gruber says. "It's pronounceable and nonoffensive and had good connotations in all languages that we've seen, and that's one of the reasons Apple kept it, I think, because it's just a good name."

According to Kittlaus, Siri, which apparently means "beautiful victorious counselor" in Norwegian, was the name he wanted to give his daughter. But he had a son, so instead, Siri was born. So what is it—not *she*; Siri is definitely not a she, or a *he*—that was getting born?

"You can think of it any way you want, but basically it's not human. If you look at the lines, like 'What's your favorite color?' And it goes . . . 'You can't see it in your spectrum' or something like that. It would be kind of like what an AI would do if you made one, it didn't grow up with a body—it has a different set of sensors. So it's kind of it's like it's trying to explain to mere mortals what it knows."

AI, which was born as a fictional conceit, has become embodied as one too.

In 2010, after they'd settled on the name and with the speech-recognition technology ready for prime time, they launched the app. It was an immediate success. "That was pretty cool," Gruber says.

"We saw it touched a nerve when we launched in the App Store as a start-up and hit top of its category in one day."

It did not take long for Apple to come knocking. "We got called real quickly after that from Apple," Gruber says. That call came directly from Steve Jobs himself. Siri was one of the final acquisitions he would oversee before his death. Apple snapped up the app for a reported $200 million—about as much as DARPA spent on the entire five-year CALO program that laid its groundwork.

At first, Siri was notorious for misinterpreting commands, and the novelty of a voice-activated AI perhaps overpowered its utility. In 2014, Apple plugged Siri into a neural net, allowing it to harness machine learning techniques and deep neural networks, while retaining many of its previous techniques, and it slowly improved its performance.

★ ★ ★

So just how smart can Siri get? "There's no excuse for it not having super-powers," Gruber says. "You know, it never sleeps, it can access the internet ten times faster than you, or whatever powers that you'd want a virtual assistant to have, but it doesn't know you."

Gruber says Siri can't offer emotional intelligence—yet. He says they need to find a theory to program first. "You can't just say, oh, 'Be a better girlfriend' or 'Be a better listener.' That's not a program-mable statement. So what you can say is 'Watch behaviors of the humans' and 'Here's the things that you want to watch for that makes them happy, and here's one thing that is bad, and do things to make them happier.' AIs will do that."

Right now, Siri is limited to performing the basic functions of the devices it lives in. "It does a lot of things, but it doesn't do all the things that an assistant can do. We think of it as, well, what do peo-ple do with their Apple devices? You navigate and you play music and that's all the things that Siri is good at right now." And Gruber and company are looking carefully at the kind of queries it routinely gets—queries that now number in the two-billion-per-week range.

"If you're in the AI game, that's like Nirvana, right?" Gruber says. "So we now know a lot about what people want in life and what they want to say to a computer and what they say to an assistant.

"We don't give it to anyone outside the company—there's a strong privacy policy. So we don't even keep most of that data on the servers, if at all, for very long. . . . Speech recognition has gotten much better because we actually look at the data and run experiments on it."

He too is fully aware of Siri's shortcomings. "Right now the illusion breaks down when either you have speech-recognition issue, or you have a question that isn't a common question or a request with an uncommon way of saying it. . . . How chatty can it get? How companion-like could it really be? Who's the audience for that? Is it kids? Is it shut-ins?

"But there are certain things you see it doesn't do right now. Like, you can't say, 'Hey, Siri, don't forget my room is 404' and 'Remind me when I'm hungry to eat or when I'm thirsty to drink water.' It can't do those things. It doesn't know the world, it doesn't see the world like we do. But if it's hooked up to sensors that do, then there's no reason why it can't."

And how does Gruber want to change Siri? "My preference is that first, it needs to be more natural in the way it speaks to you." He hopes to drive Siri to behave more like us. "So I want it to be a lot more humanlike and to not beep and make all this silly 'It's your turn now' and all that. That'll come. That'll just come naturally. I am interested in focusing in on human needs. . . . That's why we did text-interface and hands-free. So there's a real genuine need for people not to be texting while driving. There's also a need for people to deal with complexity. That's hard to do right now. So Siri is kind of a GUI-buster, like it can break through all these complicated interfaces, and you can just say, 'Remind me when I get home to call my mother' and it can know when you get home, go, 'Here's a note. Click here to call your mother.' Yeah, it could know when you're home if you tell it in your address book, then it knows when you're there from GPS. It knows your mom, it knows who your mom is, it knows what her number is, all that stuff."

SIRI will then be accessing even more of our most personal data, I note. I ask him if he worries that Siri or any other AI could do something malignant with that sort of information And, I guess, generally, is the father of Siri worried at all about the dawn of true artificial intelligence?

"I'm not afraid of general intelligence in a computer," Gruber says. "It will happen and I like it. I'm looking forward to it. It's like being afraid of nuclear power—you know, if we designed nuclear technology knowing what we know now, we could make it safe, probably." Yet critics like Elon Musk and Stephen Hawking have raised concerns that AI could evolve more quickly than we could control it—that it could pose an existential threat to humanity.

"Oh, it's great," Gruber says of the discussion. "We're kind of at the stage now where the Elon Musks of the world are saying, 'Look, this is going to be powerful enough to destroy the earth. Now how do we want to deal with the technology?' And I don't like the way that we have dealt with nuclear, but it hasn't killed us. I think we can do a lot better, but we have managed to thread that gauntlet, where we made it through the Cold War and didn't kill ourselves." Anyway, we don't have to worry about Siri. "Siri wasn't really about general intelligence, it's about intelligence at the interface. So to me that's the big problem. Our intel, our interfaces are hard to use and needlessly so."

There's also plenty of room for AI to do good—which, as a matter of fact, is why Gruber's here. He'd come on the TED cruise to see if there were any ways he could harness his expertise to help benefit ocean conservation. So far, he'd met with teams to discuss using pattern-recognition software and Google Earth to catch poachers and polluters.

"Those are kind of the superpowers that only science fiction was talking about a few years ago," he says. So, I ask, does the co-creator of Siri use his own AI? How?

"Oh, yeah, all the time," he says. "I use it twenty to thirty times a day. I mean, I get up: What's the traffic? Open an app by name. I text people back and forth by Siri. Call people by name. Get in the car. Read notifications to me, respond to texts, obviously do navigation. So, the car. Find out where I'm going. Gas on the way to work. 'Siri, where is this gas station? Take me to work.' Get to work. 'Siri, what's my next meeting?' You know, 'Change my two o'clock to three o'clock.' I mean, all that stuff and just all day long."

And then, I figure, it's time for the million-dollar question: "Do you know Siri better than it knows you? Or does Siri know you better?"

"That's a fun question. I'm afraid we're in the phase of the technology where I know Siri better than it knows me," Gruber says. "But I'd like to turn that table around soon."

CHAPTER 11

A Secure Enclave

What happens when the black mirror gets hacked

Half an hour after I showed up at Def Con, my iPhone got hacked.

The first rule of attending the largest hacker conference in North America is to disable Wi-Fi and Bluetooth on all your devices. I had done neither. Soon, my phone had joined a public Wi-Fi network, without my permission.

I had trouble with Safari when I tried to use Google; instead of search results, the page froze in the process of, it seemed, loading another page altogether.

The good thing about getting hacked at Def Con, though, is that you are surrounded by thousands of information-security pros, most of whom will happily and eloquently tell you exactly how you got "pwned."

"You probably got Pineapple'd," Ronnie Tokazowski, a security engineer for the West Virginia cybersecurity company PhishMe, tells me at the kind of absurd, faux-outdoors, French-themed buffet you can find only in a Las Vegas casino. We're joined by veteran hacker (and magician) Terry Nowles and a father and son from Minnesota; dad's a dentist, Don's into Def Con.

"The way the Wi-Fi Pineapple works is whenever your phone sends a beacon to look for an access point, instead of the Wi-Fi point saying, 'I'm that connection,' the wireless Pineapple will say, 'Yes, that's me, go ahead and connect,'" Tokazowski said. "Once you're connected to the Pineapple, they can then mill your connection, they can reroute your traffic elsewhere, they can break your traffic, they can sniff passwords."

"They can see what I'm doing on my phone, basically," I say.

"Yeah."

"Could they then actually change anything on my phone?"

"They would be able to sniff the traffic," he says, meaning intercept the data passing through the network. "Once you're connected to the network, they could start trying to throw attacks at your phone…But for the most part, the Pineapple is more for sniffing traffic." If I logged on to Gmail, for instance, the hackers could force me to go somewhere else, a site of their choosing. Then they could launch a man-in-the-middle attack. "If you went to Facebook and went to your bank account, they'd be able to see that information too," he says. "So, yeah, you just want to be careful not to connect to any Wi-Fi."

Okay, but how common is this, really?

"Pineapples?" Ronnie says. "I can go buy one for a hundred, a hundred twenty bucks. They're very, very, very common. Especially here."

Def Con is one of the largest and most notorious hacker gatherings in the world. For one weekend a year, twenty thousand hackers descend on Las Vegas to attend talks from the field's luminaries, catch up with their contemporaries, bone up on the latest exploits and system vulnerabilities, and hack the shit out of one another.

It's also one of the best places to head if you want to wade into the security issues that confront iPhones and iPhone users the world over. As more people start regarding smartphones as their primary internet devices and conducting more of their sensitive affairs on them, smartphones are increasingly going to become targets of hackers, identity thieves, and incensed ex-lovers.

Earlier, Def Con's sister conference, the smaller, more expensive, and more corporate-friendly Black Hat, had made a surprise announcement that Apple's head of security engineering and architecture, Ivan Krstić, would give a rare public talk about iOS security.

* * *

In December 2015, Syed Rizwan Farook and Tashfeen Malik, a married couple who say they were acting on behalf of ISIS, shot and killed fourteen people and seriously wounded twenty-two more at a Christmas party at the San Bernardino County Department of Public Health, where Farook worked. The spree was declared an act of terror and was, at the time, the worst domestic attack since 9/11.

During the FBI's investigation, the agency recovered an iPhone 5c. It was owned by the county—and thus was public property—but it was issued to Farook, who had locked it with a personal passcode. The FBI couldn't open the phone.

You probably have a passcode on your phone (and if you're one of the 34 percent of smartphone users who don't use a password, you should!), ranging from four numbers (weak) to the new default of six characters or longer. If you input the wrong code, the screen will do that shake/buzz thing that sort of resembles a torpedo hit in old sci-fi movies. Then it makes you wait eighty milliseconds before trying again. Every time you get it wrong, the software forces you to wait longer before your next attempt, until you're locked out completely.

For hackers, there are two main ways to break through a password. The first is via social engineering—watching (or "sniffing") a mark to gather enough information to guess it. The second is "brute-forcing" it—methodically guessing every single code combination until you hit the right one. Hackers—and security agencies—use sophisticated software to pull this off, but it can nonetheless take ages. (Imagine fiddling through every potential combination on a Master Lock.) With Farook dead—he was killed in a shootout with police—the FBI had to brute-force the phone.

But the iPhone is designed to resist brute-force attempts, and newer models eventually delete the encryption key altogether, rendering the data inaccessible. So the FBI needed a different way around this. First, they asked the National Security Agency to break into the phone. When the NSA couldn't, they asked Apple to open it for them. Apple refused and eventually issued a straightforward public response, essentially saying, *We couldn't do it even if we wanted to. And we don't want to.*

The company says it designs iPhone hardware and software to prioritize user security and privacy, and many cybersecurity experts agree that it's one of the most secure devices on the market. One reason for this is that Apple doesn't know your personal passcode—it's stored on the phone itself, in an area called the Secure Enclave, and paired with an ID number specific to your iPhone.

This maximizes consumer security but is also a proactive maneuver against federal agencies, like the FBI and the NSA, that push tech companies to install back doors (ways to covertly access user data) in their products. The documents leaked by ex-NSA whistleblower Edward Snowden reveal that the NSA has pressed major tech companies to participate in programs, like PRISM, that allow the agency to request access to user data. The documents also indicate that, as of 2012, Apple (along with Google, Microsoft, Facebook, Yahoo, and other tech companies) had been participating, though the company denies it.

So when the FBI asked Apple to give them access to Farook's phone, the company couldn't just hand over the passcode. *But.* The code that enables the time delays between failed password attempts is a part of the iPhone operating system. So the FBI made an extraordinary—and maybe unprecedented—demand: they told Apple to hack its own marquee product so they could break into the killer's phone. The FBI's court order required Apple to write new software, basically creating a custom version of its iOS—a program security experts took to calling FBiOS—that would override the delay system that prevented brute-force attacks.

Apple refused, saying that the request was an unreasonable burden and would set a dangerous precedent. The feds disagreed, arguing that Apple wrote code for its products all the time, so why not just help unlock a terrorist's cell phone?

The clash made headlines around the world. Security experts and civil libertarians praised Apple for protecting its consumers even when it was deeply unpopular to do so, while hawks and public opinion turned against the company.

Regardless, the episode gave rise to a number of pressing questions increasingly being asked of a society conducted on smartphones: How secure should our devices be? Should they be impenetrable to anyone but the user? Are there circumstances when the government should be able to gain access to a citizen's private data—like, when that citizen is a known mass murderer? That's an extreme example. But authorities are pursuing less sensational use cases too; take the NSA's routine surveillance of cell phone metadata, for example, or police departments proposing a system that would enable them to open drivers' smartphones if they've been spotted texting and driving.

This is a glitchy paradox of the moment. We share more information than ever across social networks and messaging platforms, and our phones collect more data about us than any mainstream device before—location data, fingerprints, payment info, and private pictures and files. But we have the same, or stronger, expectations of privacy as people did in generations past.

So, to keep safe the things that hackers might want most—bank-account info, passwords—Apple designed the Secure Enclave.

"We want the user's secrets to not be exposed to Apple at any point," Krstić told a packed house in Mandalay Bay Casino in Las Vegas. The Secure Enclave is "protected by a strong cryptographic master key from the user's passcode...offline attack is not possible."

And what, pray tell, does it do?

Dan Riccio, senior vice president of hardware engineering at Apple, explained it thusly when he first introduced the chip to the public: "All fingerprint information is encrypted and stored inside a secure enclave in our new A7 chip. Here it's locked away from everything else, accessible only by the Touch ID sensor. It's never available to other software, and it's never stored on Apple servers or backed up to iCloud." Basically, the enclave is a brand-new subcomputer built specifically to handle encryption and privacy without ever involving Apple's servers. It's designed to interface with your iPhone in such a way that your most crucial data stays private and entirely inaccessible to Apple, the government, or anyone else.

Or, in Krstić's words: "We can emit secret data into a page that the process can execute but that we cannot read." The enclave automatically encrypts the data that enters it, and that includes data from the Touch ID sensor.

So why do we need all these layers of extra protection? Can't Apple trust users to safeguard their own data?

"Users tend to not choose cryptographically strong passwords," Krstić said. A more aggressive quotation appeared on the screen behind him: "Humans are incapable of securely storing high quality cryptography." At the end of his talk, I was still curious about what sort of security issues Apple deals with on a regular basis. Krstić was hosting a Q-and-A, but I was fully aware that nobody keeps a high-level job in Cupertino without being skilled at some class-A evasion. Still, I had to try. I walked up the center aisle to the mike.

"What are the most persistent security issues Apple faces in iOS?" I asked.

"Tough audience questions," he replied after a moment of silence. The crowd went wild—or at least as wild as an auditorium filled with enterprise info-sec professionals at three o'clock on the last day of a conference could—with applause and laughter. "Thank you," he said as I waited for an answer. "No—thank you," he repeated. That was all the answer I was going to get.

* * *

The FBI's efforts to hack into the iPhone may have drawn cybersecurity into the spotlight, but hackers have been cracking the device since the first day of its launch. As with most other modern electronics, hacking has helped shape the culture and contours of the products themselves. It boasts a storied and slightly noble legacy;

hacking has been around for as long as people have been transmitting information electronically.

One of the first and most amusing historical hacks was launched in 1903 over a wireless network. The Italian radio entrepreneur Guglielmo Marconi had organized a public demonstration of his brandnew wireless-communications network, which, as he had boldly announced, could transmit Morse code messages over great distances. And, he claimed, it could do so entirely securely. He said that by tuning his apparatus to a specific wavelength, only the intended party could receive the message being sent.

His associate Sir John Ambrose Fleming set up a receiver in the Royal Institution's lecture hall in London; Marconi would transmit a message to it from a hilltop station three hundred miles away in Poldhu, Cornwall. As the time for the demonstration grew near, a strange, rhythmic tapping sound became audible. It was Morse code, and someone was beaming it into the lecture hall. At first, it was the same word repeated over and over: *Rats*. Then the sender got poetic with a limerick that began *There was a young fellow of Italy, who diddled the public quite prettily*. Marconi and Fleming had been hacked.

A magician named Nevil Maskelyne declared himself the culprit. He'd been hired by the Eastern Telegraph Company, which stood to lose a fortune if someone found a way to transmit messages that was cheaper than the company's terrestrial networks. After Marconi announced his secure wireless line, Maskelyne built a one-hundred-and-fifty-foot radio mast near the transmission route to see if he could eavesdrop on it. Marconi's system was, in hindsight, anything but secure. His patented technology that allowed him to tune his transmission to a specific wavelength is now essentially what a radio station does to broadcast its programs to all of the public; if you have the wavelength, you can listen in.

When Maskelyne demonstrated that fact to the audience at the lecture hall, the public learned of a major security flaw in new technology, and Maskelyne enjoyed some of the first lulz.

Hacking as the techno-cultural phenomenon that we know today probably picked up steam with the counterculture-friendly phone phreaks of the 1960s. At the time, long-distance calls were signaled in AT&T's computer routing system with a certain pitch, which meant that mimicking that pitch could open the system. One of the first phone phreaks was Joe Engressia, a seven-year-old blind boy with perfect pitch (he'd later rename himself Joybubbles). He discovered that he could whistle at a certain frequency into his home phone and gain access to the long-distance operator, for free. John Draper, another legendary hacker who came to be known as Captain Crunch, found that the pitch of a toy whistle that came free in Cap'n Crunch cereal boxes could be used to open long-distance call lines; he built blue boxes, electronic devices that generated the tone, and demonstrated the technology to a young Steve Wozniak and his friend Steve Jobs. Jobs famously turned the blue boxes into his first ad hoc entrepreneurial effort; Woz built them, and Jobs sold them.

The culture of hacking, reshaping, and bending consumer technologies to one's personal will is as old as the history of those technologies. The iPhone is not immune. In fact, hackers helped push the phone toward adopting its most successful feature, the App Store.

* * *

The fact that the first iPhones were sold exclusively through AT&T meant that they were, in a sense, a luxury phone. At $499 for the low-end 4G model, they were expensive. Every Apple diehard around the world wanted one immediately, but unless you were willing to sign on with AT&T and you lived in the United States, you were out of luck.

It took a seventeen-year-old hacker from New Jersey a few weeks to change that.

"Hi, everyone, this is Geohot. And this is the world's first unlocked iPhone," George Hotz announced in a YouTube video that was uploaded in July 2007. It's since been viewed over two million times. Working with a team of online hackers intent on freeing the iPhone from its AT&T bondage, Hotz logged five hundred hours investigating the phone's weaknesses before finding a road map to the holy grail.

He used an eyeglass screwdriver and a guitar pick to remove the phone's back and found the baseband processor, the chip that locked the phone onto AT&T networks. Then he overrode that chip by soldering a wire to it and running enough voltage through it to scramble its code. On his PC, he wrote a program that enabled the iPhone to work on any wireless carrier.

He filmed the result—placing a call with an iPhone using a T-Mobile SIM card—and shot to fame. A wealthy entrepreneur traded him a sports car for the unlocked phone. Apple's stock price rose on the day the news broke, and analysts attributed that to the fact that people had heard you could get the Jesus phone without AT&T.

Meanwhile, a group of veteran hackers calling themselves the iPhone Dev Team had organized a break into the iPhone's walled garden.

"Back in 2007, I was in college, and I didn't have a lot of money," David Wang says. As a gearhead, he was intrigued when the iPhone was announced. "I thought it was a really impressive, important milestone for a device—I really wanted it." But the iPhone was too expensive for him, and you had to buy it with AT&T. "But they also announced the iPod Touch, and I was like, I can afford that…I thought, you know, I could buy an iPod Touch, and they'll eventually release a capability to let it make web calls, right?"

Or he could just try to hack it into one.

"At the time, there was no App Store, there was no third-party apps at all," Wang says. "I was hearing stuff about people who were modding it, the iPhone Dev Team, and the hackers, and how they got code execution on the iPhone. I was waiting for them to do the same with iPod Touch."

The iPhone Dev Team was perhaps the most prominent hacker collective to take aim at the iPhone. They started probing the phone for vulnerabilities in its code, bugs they would be able to exploit to take over the phone's operating system. Wang was watching, and waiting.

"Every product starts out in an unknown state," the cybersecurity expert Dan Guido tells me. Guido is the co-founder of the cybersecurity firm Trail of Bits, which advises the likes of Facebook and DARPA. He was formerly an intelligence lead at the New York Federal Reserve, and he's an expert on mobile security. Apple, he says, "lacked a lot of exploit mitigations, they had lots of bugs in really critical services." But that was to be expected. It was a new frontier, and there were going to be pitfalls.

"One person found that the iPhone and the iPod Touch was vulnerable to this TIFF exploit," Wang said. A TIFF is a large file format commonly used for images by desktop publishers. When the device went to a site displaying a TIFF, Wang says, "Safari would crash, because the parser had a bug in it"—and you could take control of the entire OS.

It took hackers only a day or two to break into the iPhone's software. Hackers would post proof of pwning the system—uploading a video of the phone with an unauthorized ringtone, for example—and then typically follow up with set of how-to instructions so other hackers could replicate it.

"When [the iPhone] came out, it was just for Mac," Wang says. In 2007, the Mac's market share was still relatively small, just 8 percent of the U.S. market. Remember the iPod lesson: Restricting users to Mac limits the audience. "I didn't want to wait for people to come up with Windows instructions, so I figured out how they were doing it, and made a set of instructions for Windows users...it turned out to be seventy-six steps." That was a turning point. Wang, whose handle is planetbeing, posted his instructions online, and it set off a frenzy. "So if you Google *seventy-six-step jailbreak*, you would see my name. It was the first thing that I did."

Jailbreaking became the popular term for knocking down the iPhone's security system and allowing users to treat the device as an actual personal computer—letting them modify settings, install new apps, and so forth. But breaking in was only the first step. "After you do that, you still have to do a lot, like install the installer app that enables you to easily install applications and all the tools, and set the root file system, read/write, and all those things, and so my steps were to help you do that. So I wrote a tool for that," Wang says.

Hacking is a competitive sport. Collectives function a bit like pro teams; you can't just show up with a ball and expect to play. Hackers have to prove themselves. "They were pretty closed off," Wang says. "There's a problem with the hacking community — they didn't want to share their techniques, [were] annoyed by kids like me who had a little skill and wanted to learn. But once you do something awesome, they let you in."

Shortly after uploading the jailbreak instructions, Wang saw a blog post by the security expert H. D. Moore, who'd taken apart, step by step, that TIFF exploit. Moore had, in essence, laid out a blueprint for an automatic jailbreak. Wang wrote the predecessor of what would become perhaps the most legendary iPhone jailbreak mechanism, an online app you could access on Safari that would immediately jailbreak the phone. Fellow Dev Team member Comex, aka Nicholas Allegra, built the actual JailbreakMe app.

"Some of the exploits that came out, like the JailbreakMe attack," were really fun, Guido says. At the time you could go into an Apple Store, open up JailBreakMe.com on a display phone, hit its Swipe to Unlock button, and "it would run the exploit and root the phone from the internet," Guido says. The Swipe to Unlock was a play on the iPhone's famous opening mechanism, a double entendre highlighting the fact that you were being freed from a closed, locked system by the Dev Team. "And you could just go to an Apple Store and jailbreak every single phone they had on display."

That's exactly what in-the-know hackers did. "A lot of people started doing that," Wang says, "because suddenly it was really, really easy."

★ ★ ★

Apple, aware that jailbreaking was becoming an increasingly mainstream trend, broke its silence on the practice on September 24, 2007, and issued a statement: "Apple has discovered that many of the unauthorized iPhone unlocking programs available on the internet cause irreparable damage to the iPhone's software, which will likely result in the modified iPhone becoming permanently inoperable when a future Apple-supplied iPhone software update is installed."

There were genuine reasons that Apple was concerned about jailbreaking. Guido says that the JailbreakMe episode "could have been turned around really quickly into an attack tool kit and we're lucky that it wasn't."

The vast majority of the jailbreakers, like Wang, were eager to expand the capabilities of a clearly capable machine. The majority weren't hacking into other people's phones (besides jailbreaking Apple Store display models, an easily reversible prank) and they were only jailbreaking their own to customize and open them up—and, of course, for the sport of it.

Apple's threat went unheeded. Apple patched the bug that enabled the TIFF exploit, setting off what would be a years-long battle. The iPhone Dev Team and other jailbreaking crews would find a new vulnerability and release new jailbreaks. The first to find a new one would get cred. Then Apple would fix the bug and brick the jailbroken phones. When asked about jailbreaking at a press event, Steve Jobs called it "a cat and mouse game" between Apple and the hackers. "I'm not sure if we are the cat or the mouse. People will try to break in, and it's our job to stop them breaking in."

Over time, the jailbreaking community grew in size and stature. The Dev Team reverse-engineered the phone's operating system to allow it to run third-party apps. Hacker-developers made games, voice apps, and tools to change the look of the phone's interface. On Apple's phone, you could customize very little—the original iPhone didn't even have an option for wallpaper; the apps just hovered on a black background. And the fonts, layout, and animations were all set in stone. It was the hackers who were pushing the device to become more like the creativity augmenter, the knowledge manipulator that Steve Jobs's idol Alan Kay originally imagined mobile computing could be.

One of the Dev Team members, Jay Freeman, or saurik, built Cydia—basically, a predecessor of the App Store accessible only on a jailbroken iPhone—and in February 2008 he released it. Cydia allowed users to do a lot more than the current App Store does; they could download apps, games, and programs, sure. But they could also download tweaks and more drastic overhauls; you could, for instance, redesign the layout of your home screen, download ad-blockers, and apps to make non-AT&T calls, and exert more control over data storage.

The popularity of jailbreaking and Cydia provided a public demonstration of a palpable demand for, at the very least, a way to get new apps, and, at the most, a way to have more control over the device. Before long, Apple declared jailbreaking unlawful, though it never actually sued any of the jailbreakers. The internet freedom advocacy group Electronic Frontier Foundation lobbied to have the practice listed as an exemption in the Digital Millennium Copyright Act, a request that a federal appeals court granted, thus closing the issue. Tim Wu, a law professor at Columbia University, famously said that "jailbreaking Apple's superphone is legal, ethical, and just plain fun."

"It's an interesting gray area, the sort we rarely see anymore—it turns out that this kind of hacking was entirely legal," Guido says. "Anyone could jailbreak their phone."

Freeman saw it as more of an ideological imperative, however. "The whole point is to fight against the corporate overlord," he told the *Washington Post* in 2011. "This is a grass-roots movement, and that's what makes Cydia so interesting. Apple is this ivory tower, a controlled experience, and the thing that really brought people into jailbreaking is that it makes the experience theirs." As of 2011, he said, his platform had 4.5 million weekly users and was generating $250,000 in revenue a year, most of which was pumped back into supporting the electronic ecosystem.

Money was an issue for the jailbreakers like iPhone Dev Team, who relied on PayPal donations and outside jobs to fund their efforts,

Wang says. Over time, as the App Store drained some of the interest in jailbreaking and as Apple became increasingly aggressive in its efforts to prevent and discourage breaks, the original team began to drift off.

And it turned out that, as with any good underground-rebels-versus-authority story, there was a twist: One of the core iPhone Dev Team members was an Apple employee. None of the Dev Team had any clue that the hacker who went by the name bushing and who was known for his skills with reverse-engineering was working for the company whose phones they were hacking.

Who was bushing? Ben Byer, who had signed on as a senior embedded security engineer with Apple in 2006. At least, that's what a web of his online trail suggests. A LinkedIn profile for Ben B. lists that job title as well as a work history that includes a stint with Libsecondlife—an effort to create an open-source version of the once-popular Second Life game, where bushing was a frequent poster.

"We didn't know it at the time," Wang says today. "We didn't realize until later...he kind of came out to us later on." Bushing would go on to be a formidable force in the hacking community. Tragically, he passed away in 2016 at the age of thirty-six due to what his friends and peers describe as natural causes.

* * *

While jailbreaking is not as sensational a practice as it once was—like any worthy tech endeavor, it's been declared dead by the pundits multiple times—the legacy of the jailbreakers remains.

"The most obvious example of Apple copying the jailbreak community is the introduction of Notification Center," Alex Heath, who now reports for *Business Insider*, wrote in 2011. He was referring to Apple's newly released notification system, which let users view a compendium of updates and messages in a single screen. "A new method of notifications has been something that iOS has needed desperately for years, and the jailbreak community has been offering alternative systems for a long time." He noted that Apple actually hired the developer of a Cydia notifications app to help build it, and visually, the systems do look similar.

Perhaps more than anything, though, the jailbreakers demonstrated living, coded proof that there was immense demand for an App Store and that people would be able to do great things with it. Through their illicit innovation, they showed that the iPhone could become a vibrant, diverse ecosystem for doing more than making calls, surfing the web, and increasing productivity. And they showed that developers would be willing to go to great lengths to participate on the platform; and they didn't just talk, they built a working model.

Thus, the hacker iPhone Dev Team should get a share of at least some of the credit in Jobs's decision to let the real iPhone Dev Team open the device to developers in 2008.

"I don't want to have too much hubris in our role. We didn't know how much Apple had planned before us," Wang says, or how much it mattered that they relentlessly hacked the iPhone until it opened up. "I want to say it does."

★ ★ ★

Another legacy of the jailbreaking movement was that it drove Apple to focus on security with renewed vigor.

"Consumers shouldn't have to think about security," Dan Guido tells me. "Apple's done extremely well at what I call 'security paternalism,'" he says. "Being the dad and telling kids they can't do things, but, for their own benefit." That's a good way to describe Apple's approach.

"They went through a really aggressive, top-down hardening campaign for the entire iOS platform over the last few years," Guido says, "and instead of thinking about it from a tactical perspective, of, like, 'Let's just fix all the bugs,' they came at it from a really architectural perspective and thought about the attacks they were gonna face and kind of predicted where some of them were going." They stopped playing cat-and-mouse with hackers and started rewriting the rules, setting out mousetraps long before the mice had a chance to sneak into the house.

Per Apple's longstanding MO, how exactly it has protected user privacy and how exactly the Secure Enclave works has been shrouded in secrecy. "An effect of this security paternalism is that if you want to investigate how secure the platform is, you can't," Guido says. Nobody outside Apple knows for sure how the device works, just that it seems to. Really well. And it's a good thing that Apple started upping its security game. "You've got heads of state walking around with iPhones," he says. "And you've got a billion sold, so you've got to assume that people are screwing around. And we have seen attacks on iPhones that don't abuse jailbreaks. They're rare."

The iPhone has been helped on this front, somewhat ironically, by the rise of Android phones. The iPhone may be the single most popular and profitable device on the planet, but it's the only phone running the iOS operating system. Samsung, LG, Huawei, and other handset manufacturers all run Android. That gives Android around 80 percent of the mobile OS market share worldwide. And malicious hackers tend to try to maximize their time and effort; for them, it's a numbers game.

"Don't try to hack the iPhone, it's too hard, you won't get anything out of it," Guido says. That's the attitude of most black-hat hackers. "Apple can smack you down really quickly. They issue patches that people actually apply." You know when Apple asks you to update your iOS? And you just *Sure, whatever,* click? Well, that patches up the most recent bugs that were exposing your phone to outside hackers and nullifies the malignant software hackers might have been trying to use to get access to your phone. And iPhone users update their phones in much larger proportions than Android users do.

Apple's more stringent app-approval process helps too. "If you do Android apps, they're so malicious," Guido says. "But on iPhone, the rigor that goes into the approval process prevents a lot of that. And Apple can disinfect remotely every phone that's infected.

As a result, the security on iPhones today is, for the most part, really good.

The iOS devices are the single most secure consumer devices available, according to Guido. "They are built like a tank from a security perspective," Guido says. "It is light-years ahead of every other trusted device that exists on the market. It has really been designed well by people who know what's going on, to keep and hold your secrets in a way that even the most well-resourced adversary can't get access."

But it's still not perfect; iPhones have nonetheless been subject to a number of high-profile hacks. Charlie Miller famously managed to get the App Store to approve a malware app that allowed him to break Apple's stranglehold on the device. For five hundred dollars, University of Michigan professor Anil Jain was able to build a device that fooled the iPhone's fingerprint sensors.

In 2015, the security firm Zerodium paid a bounty of one million dollars for a chain of zero-day exploits (vulnerabilities that the vendor isn't aware of) on the iPhone, though no one knows who won the money. And no one, save Zerodium, knows what became of the zero days. And in 2016, Toronto's Citizen Lab revealed that a very sophisticated form of malware, called Trident, had been used to try to infect a civil rights activist's phone in the UAE. The hack was revealed to have been the work of an Israeli company, which was believed to have sold its spyware for as much as $500,000—likely to authoritarian regimes like the UAE government.

The majority of those hacks are unlikely to affect most users. "You've got to look at the bigger picture: More and more people are using non-general-purpose computing devices, they're using Kindles, iPads, ChromeBooks, iPhones, Apple TV, whatever, all these locked-down devices that serve one single purpose," Guido says. "And it's significantly harder to get malware on those because they're not general purpose. I think the world is shifting. Not just Apple. General-purpose computers are taking less of a primary role in our lives, and it's going to pay off tremendously well for security."

Even the best-secured devices aren't perfect, and locked-down, single-purpose devices are definitely vulnerable to attacks, especially since they are all increasingly connecting to the internet. I can tell you from experience. Yep, the same hack that snared my iPhone.

"Wi-Fi attacks have strangely not gone away," Guido says. "They're one of these unsexy problems that people just don't seem interested in solving. If someone really wants to exploit that if they put you on a Wi-Fi network and want to gain access to your phone, there are certain low-resource attacks they can do—they can try to redirect you to another website when you open up Safari and try to convince you to put your password in somewhere. But that's a little intrusive—you get caught that way."

Basically, these rules apply whenever you use public Wi-Fi—don't enter any sensitive data over public networks and log in to only those networks that you trust. Update your phone when prompted.

The landscape is changing—as Guido noted, there are more persons of interest with iPhones and more of an imperative to hack into those phones. It's less likely to be done by loose-knit hackers out for the lulz or to earn a few bucks; it's more likely to come from a government agency or a well-paid firm that does business with government agencies. Security pros are skeptical when the FBI says it needs back doors to combat ISIS and track encrypted recruitment and terrorist-plotting efforts, because its inability to prevent such attacks is so evident. But there are other situations—such as when photos that Apple helped law enforcement unlock sent two people who had sexually abused a sixteen-month-old child to prison—that help make a case for Apple's cooperation. (Which, it should be added, the company has provided in the past: Apple has reportedly opened over seventy iPhones at the behest of law enforcement, though many of those were before the Secure Enclave necessitated a novel software hack from Apple.) There may need to be a mechanism for law enforcement to access this stuff, but how we do that in the age of the Secure Enclave is an open question.

For Apple, security is a question of product too. As it moves to promote Apple Pay, internet-of-things apps, and HealthKit, consumers must be confident their data can be kept safe. From a consumer's perspective, Apple's decision is win-win; it may be unpopular, but the message is clear: You won't find a more secure phone anywhere. We'll go to bat against the feds to make sure your phone is secure. Even if you're a terrorist, your data is safe.

Looking for a little more clarity, after Apple's security guru was done with his talk, I walked over behind the stage, where a small crowd was gathering. I asked him how he felt the cybersecurity scene was changing with the dominance of smartphones.

"Well, one part of the landscape that is changing is—"

"So the PR guy is going to jump in," the Apple PR guy said, actually jumping in, thrusting a card into my hand, and shepherding Krstić away.

Of course, Apple was going to keep its Secure Enclave secret.

CHAPTER 12

Designed in California, Made in China

The cost of assembling the planet's most profitable product

All gray dormitories and weather-beaten warehouses, the sprawling factory compound blends seamlessly into the outskirts of the Shenzhen megalopolis. Foxconn's enormous Longhua plant is a major manufacturer of Apple products; it might be the best-known factory in the world. It might also might be among the most secretive and sealed-off. Security guards man each of the entry points. Employees can't get in without swiping an ID card; drivers entering with delivery trucks are subject to fingerprint scanners. A Reuters journalist was once dragged out of a car and beaten for taking photos from outside the factory walls. The warning signs outside—THIS FACTORY AREA IS LEGALLY ESTABLISHED WITH STATE APPROVAL. TRESPASSING IS PROHIBITED. OFFENDERS WILL BE SENT TO POLICE FOR PROSECUTION!—are more aggressive than those outside many Chinese military compounds.

But it turns out that there's a secret way into the heart of the infamous operation: Use the bathroom. I couldn't believe it. Thanks to a simple twist of fate and some clever perseverance by my fixer, I'd found myself deep inside so-called Foxconn City.

★ ★ ★

It's printed on the back of every iPhone: DESIGNED IN CALIFORNIA BY APPLE, ASSEMBLED IN CHINA. U.S. law dictates that products manufactured in China must be labeled as such, and Apple's inclusion of the *designed by* phrase renders the statement uniquely illustrative of one of the planet's starkest economic divides. The cutting edge is conceived and designed in Silicon Valley, but it is assembled by hand in China.

The vast majority of plants that produce the iPhone's component parts and carry out the device's final assembly are based here, in the People's Republic, where low labor costs and a massive, highly skilled workforce have made the nation the ideal place to manufacture iPhones (and just about every other gadget). The country's vast, unprecedented production capabilities—the U.S. Bureau of Labor Statistics estimated that as of 2009 there were ninety-nine million factory workers in China—has helped the nation become the world's largest economy. And since the first iPhone shipped, the company doing the lion's share of the manufacturing is the Taiwanese Hon Hai Precision Industry Company, Ltd., better known by its trade name, Foxconn.

Foxconn is the single largest employer on mainland China; there are 1.3 million people on its payroll. Worldwide, among corporations, only Walmart and McDonald's employ more. As of 2016, that was more than twice as many people working for the five most valuable tech companies in the United States—Apple (66,000), Alphabet (née Google, 70,000), Amazon (270,000), Microsoft (64,000), and Facebook (16,000)—combined. More people work for Foxconn than live in Estonia.

Today, the iPhone is made at a number of different factories around China, but for years, as it became the bestselling product in the world, it was largely assembled at Foxconn's 1.4-square-mile flagship plant here, just outside of the manufacturing megalopolis of Shenzhen. The sprawling factory was once home to an estimated 450,000 workers. Today, that number is believed to be smaller, but it remains one of the biggest such operations in the world.

If you know of Foxconn, there's a good chance it's because you've heard of the suicides. In 2010, Longhua assembly-line workers began committing suicide en masse. Worker after worker threw him- or herself off the towering dorm buildings, sometimes in broad daylight, in tragic displays of desperation—and in protest of the work conditions inside. There were eighteen reported suicide attempts that year alone, and fourteen confirmed deaths. Twenty more workers were talked down by Foxconn officials.

The epidemic caused a media sensation—suicides and sweatshop conditions in the House of iPhone. Suicide notes and survivors told of immense stress, long workdays, and harsh managers who were prone to humiliate workers for mistakes; of unfair fines and unkept promises of benefits.

The corporate response spurred further unease: Foxconn CEO Terry Gou had large nets installed outside many of the buildings to catch falling bodies. The company hired counselors, and workers were made to sign pledges stating they would not attempt suicide. Commentators suggested that a lot of the suicides were migrant workers who had trouble adjusting to the rapid-fire pace of urban environs. Steve Jobs, for his part, declared, "We're all over that" when asked about the spate of deaths, and he pointed out that the rate of suicides at Foxconn was within the national average and lower than at many U.S. universities. Critics pounced on the comments as callous, though he wasn't technically wrong. Foxconn Longhua was so massive that it could be its own nation-state, and the suicide rate was comparable to its host country's. The difference is that Foxconn City is a nation-state governed entirely by a corporation, and one that happened to be producing one of the most profitable products on the planet.

Since 2010, Foxconn and Longhua have been in and out of the media spotlight, though poor conditions, worker unrest, and even suicides have continued. Meanwhile, Apple's other major iPhone manufacturer, the Shanghai-based Pegatron, Foxconn's rival, has been charged with exploiting workers and forcing brutal stretches of overtime in patterns that eerily mimic its competition's. An investigation revealed that workers were routinely logging hundred-hour workweeks and toiling as many as eighteen days in a row, and the

BBC obtained footage of workers falling asleep on the assembly line. Labor advocates worried that Pegatron was even *worse*.

So I traveled to China to try to get an up-close look at what it took to manufacture the world's most profitable product, designed by a globally celebrated innovation engine some five thousand miles across the Pacific, in a nation that's both the prime producer of the one device and its fastest-growing market. First stop, Shanghai.

★ ★ ★

Somewhere, in nearly every corner of this sprawling city, someone is building a part that will end up in an iPhone, or maybe snapping the whole thing together. Of the two hundred addresses that Apple lists for its top suppliers on its annual report, nearly half are located in just two cities: here and Shenzhen.

The forty suppliers here in Shanghai, like TSMC, the chip manufacturer that produces the iPhone's ARM-based brain, are scattered far and wide across the city.

When I arrive at TSMC's headquarters, the security checkpoint is posted far from the complex, so I can't see much of anything besides the well-groomed lawn and the mammoth gray-and-red plant walls. The guards, of course, won't let me in for a closer look. I snap some photos and jog back to the idling car. A security guard follows me, yelling. He's demanding that I delete the photos and won't let us leave until I pretend to have done so. It's a recurring theme during my sightseeing tour of Apple's suppliers. In fact, I'd soon be able to tell immediately which building in a given neighborhood housed an Apple component factory: it was the one with high security, barbed wire, or posted guards.

That was especially true of Pegatron, which had cameras loaded with facial-recognition software at the entrance; every worker, all of them forming a human river that flowed into the factory's mouth, swiped a card and glanced into the camera, and the turnstile clacked open. Pegatron's on the outskirts of the city, a subway stop away from Disney Shanghai; I walk the perimeter with my fixer and find it swarming with hundreds of college-age workers with lanyards dangling around their necks. We pass a fortune-teller, and I hand him ten renminbi to tell me the future of the iPhone. "Everyone says it is a good phone and the future is getting better because it's increasingly profitable," he says. Though he also tells me that I have a good face and that women are going to chase me around, so I'm not entirely sure he's to be trusted. We interview as many workers as we can and begin to confirm a picture of a high-stress workplace marked by long hours and repetitive tasks, a factory where most hires last only about a year before quitting.

It's no exaggeration to say that the iPhone has transformed China. On top of physically building the device, China is now one of the world's top consumer markets too. Shanghai is fascinating—a blend of enthusiastic entrepreneurship and manufacturing muscle dominated its smartphonic tech sector. But it's got nothing on Shenzhen.

* * *

Shenzhen was the first SEZ, or special economic zone, that China opened to foreign companies, beginning in 1980. At the time, it was a fishing village that was home to some twenty-five thousand people. In one of the most remarkable urban transformations in history, today, Shenzhen is China's third-largest city, home to towering skyscrapers, millions of residents, and, of course, sprawling factories. And it pulled off the feat in part by becoming the world's gadget factory. An estimated 90 percent of the world's consumer electronics pass through Shenzhen.

Just across from Hong Kong, on China's mainland, downtown Shenzhen feels sleek, new, strained, and chaotic. Traffic is snarled and the lights beat neon, but Shenzhen often seems more mall-punk than cyberpunk.

"I believe Shenzhen embodies the spirit of China," says Isaac Chen, who was born in Shenzhen after his parents moved there in the 1990s in the first wave of the business boom and whom I was fortunate enough to sit next to on the plane. "People working very hard, long hours, in new industries. I was among the first generation to be born there," he says. "When I was a kid, there were hills everywhere. Now it is flat. They destroyed the hills to build the coastline. It is completely changed."

Chen says the conditions in many factories there are "brutal," though he does not say this sorrowfully. "When we were in Paris, we met a sweeper; he spent all day sweeping the same road, and took pride in the fact that he had done the job well for twenty years. We could not understand this. In China, we always want to improve. There is a fear that if we do not, we will have to go back to nothing, back to farming the land for food," he says. "China is all about work. Work and money. We do not take vacations."

* * *

A cabdriver let us out in front of the factory; boxy blue letters spelled out FOXCONN next to the entrance. It was a typically gray day in Shenzhen. The security guards eye us, half bored, half suspicious. My fixer, a journalist from Shanghai who I'll call Wang Yang, and I

A main entrance to Foxconn's Longhua factory.

decide to walk the premises first and talk to workers, to see if there might be a way to get inside.

The first people we stop turn out to be a pair of former Foxconn workers. Neither is shy.

"It's not a good place for human beings," says one of the young men, who goes by Xu. He'd worked in Longhua for about a year, until a couple months ago, and he says the conditions inside are as bad as ever. "There is no improvement since the media coverage," Xu says. The work is very high pressure, and he and his colleagues regularly log twelve-hour shifts. Management is both aggressive and duplicitous, publicly scolding workers for being too slow and making them promises they don't keep, he says. His friend, who worked at the factory for two years and chooses to stay anonymous, says he was promised double pay for overtime hours but got only regular pay. He says he was promised a raise but never received it. "So that's why we wanted to leave the company."

They paint a bleak picture of a high-pressure working environment where exploitation is routine, and where depression and suicide have become normalized.

"It wouldn't be Foxconn without people dying," Xu says. "Every year people kill themselves. They take it as a normal thing."

* * *

Over several visits to different iPhone assembly factories in Shenzhen and Shanghai, we interviewed dozens of such workers. Let's be honest: To get a truly representative sample of life at an iPhone factory would require a massive canvassing effort and the systematic and clandestine interviewing of thousands of employees. So take this for what it is—efforts to talk to often skittish, often wary, and often bored workers who were coming out of the factory gates, taking a lunch break at a nearby noodle shop, or congregating somewhere after their shifts.

The vision of life inside an iPhone factory that emerged was varied—some found the work tolerable, others were scathing in their criticisms, some personally experienced the despair Foxconn was known for, and still others had taken jobs there just to try to find a girl-friend. Most knew of the reports of poor conditions before joining, but they either needed the work or it didn't bother them. Almost every-where, people said the workforce was young, and turnover was high. "Most employees last only a year" was a common refrain.

Perhaps that's because of the pace of work is widely agreed to be relentless, and the management culture was often described as cruel.

Since the iPhone is such a compact, complex machine, putting one together correctly requires sprawling assembly lines of hundreds of people who build inspect, test, and package each device. One worker said seventeen hundred iPhones passed through her hands every day; she was in charge of wiping a special polish on the display. That comes out to polishing about three screens a minute for twelve hours a day. Another said he worked as part of an inspection team of two or three people, and they were in charge of doing quality assurance for three thousand iPhones a day.

More meticulous work, like fastening chip boards and assembling back covers, was slower; these workers have a minute apiece for each iPhone. That's still around six or seven hundred iPhones a day. Fail-ing to meet quota or making a mistake can draw a public condemna-tion from superiors. Workers are often expected to stay silent and may draw rebukes from their bosses for asking to use the restroom.

Xu and his friend were both walk-on recruits, though not neces-sarily willing ones.

"They call Foxconn a Fox Trap," he says. "Because it tricks a lot of people."

"I was tricked to work for Foxconn," Xu says. "I intended to work for Huawei," he adds, referring to the Chinese smartphone competi-tor. "People feel way better working for Huawei, better corporate culture, more comfortable." In fact, he says, "Everyone has the idea of working in Foxconn for one year and getting out of the factory and going to work for Huawei."

But when he went to a recruiting office, they told him Huawei already had enough workers and they took him to Foxconn. He believes this is because Foxconn pays recruiters extra to find more people—it simply wasn't true that Huawei was full.

And that, he says, was just the first part of the Fox Trap. "They just didn't keep their promises, and that's another way of tricking you." He says Foxconn promised them free housing but then forced them to pay exorbitantly high utility bills for electricity and water. The current dorms sleep eight to a room, and he says they used to be twelve to a room. But Foxconn would shirk on social insurance and be late or fail to pay bonuses. And many workers sign contracts that subtract a hefty penalty from their paychecks if they quit before a three-month introductory period. "We thought Foxconn was a good factory to work in, but we found out once we got there that it was not."

On top of all that, the work is grueling. "If you got one hundred salary, you have to pay three hundred effort," Xu says. "You have to have mental management"—otherwise, you can get scolded by bosses in front of your peers. Instead of discussing performance privately or face to face on the line, managers would stockpile complaints until later. "When the boss comes down to inspect the work, they get a heads-up to prepare," Xu's friend says. "If the boss finds any problems, they won't scold you then. They will scold you in front of everyone in a meeting later."

These meetings are apparently routine. At the end of the day, the manager will ask everyone on a team to stand up and gather around. In addition to praising productive workers and offering a general debriefing, the manager will single out anyone he or she believes made mistakes.

"It's insulting and humiliating to people all the time," his friend says. "Punish someone to make an example for everyone else. It's systematic," he adds. "There are bonuses, and if you get scolded you won't get the bonus."

In certain cases, if a manager decides that a worker has made an especially costly mistake, the worker has to prepare a formal apology. "They must read a promise letter aloud—'I won't make this mistake again'—to everyone." One of his colleagues, who took the blame for someone else's mistake to protect them, "cried, [he was] scolded so badly."

This culture of high-stress work, anxiety, and humiliation contributes to widespread depression.

Xu says there was another suicide a few months ago. He saw it himself. The victim was a college student who worked on the iPhone assembly line. "Somebody I knew, somebody I saw around the cafeteria," he says. After being publicly scolded by a manager, he got into a quarrel. Company officials called the police, though the worker hadn't been violent, just angry.

"He took it very personally," Xu says, "and he couldn't get through it." Three days later, he jumped out of a ninth-story window. "I was out for lunch, and saw everyone making a scene. He was on the ground surrounded in blood."

So why didn't the suicide get any media coverage? I ask. Xu and his friend look at each other and shrug. "Here someone dies, one day later the whole thing doesn't exist," his friend says. "You forget about it."

Xu Lizhi, who committed suicide at Longhua in September 2014, left behind diaries and poetry that opened a window into that attitude.

A Screw Fell to the Ground

A screw fell to the ground / In this dark night of overtime / Plunging vertically, lightly clinking / It won't attract anyone's attention / Just like last time / On a night like this / When someone plunged to the ground—9 January 2014

"We are on top of this. We look at everything at these companies," Steve Jobs said after news of the suicides broke. "Foxconn is not a sweatshop. It's a factory—but my gosh, they have restaurants and movie theaters...but it's a factory. But they've had some suicides and attempted suicides—and they have 400,000 people there. The rate is under what the US rate is, but it's still troubling." Tim Cook visited

Longhua in 2011 and reportedly met with suicide-prevention experts and top management about the epidemic.

In 2012, a hundred and fifty workers gathered on a rooftop and threatened to jump. They were promised improvements and talked down by management; they had, essentially, wielded the threat of suicide as a bargaining tool. In 2016, a smaller group did it again. Just a month before we spoke, Xu says, seven or eight workers gathered on a rooftop and threatened to jump unless they were paid the wages they were due, which had apparently been withheld. Eventually, Xu says, Foxconn agreed to pay the wages, and the workers were talked down.

Everyone has gotten used to the "ghost of death" at Foxconn. Foxconn claims that they're working on the problem, but he thinks even company officials don't know what to do. "Everyone thinks it is cursed." In addition to the nets and the counseling, administrators have tried other, more unconventional means too.

"They built a tower to scare the ghosts away," Xu says. "In any buildings that don't look 'normal,'" he says, "they keep the lights on all day for superstition."

Xu and his friend call the action of suicide "pretty silly" and say they left because of the day-to-day dehumanization. They had been approached about joining management, they say—perhaps another part of the Fox Trap—before bailing. Xu had begun training. "I couldn't bear with everything," he says. "Couldn't stand it. They forced me to do things I didn't want to do," like discipline and humiliate workers. "If you didn't obey their ways they reduced your salary." He says, with a hint of pride, that though he thought he could do the job, it wasn't worth it. He didn't want to give anyone such a hard time, he says. "Even if they offered much more salary I wouldn't take it."

All of the above is why the turnover rate is so high; there are very few longtime workers here, Xu says. "There were fifteen people with me when I entered the factory. Now there are only two left." Not including him—he quit to go to work at an electronics shop. He says that he's "absolutely more happy now that I've left the factory."

When I ask him about Apple and the iPhone, his response is swift:

"We don't blame Apple. We blame Foxconn." When I ask them if they would consider working at Foxconn again if the conditions improved, the response is equally blunt.

"You can't change anything," Xu says. "It will never change."

★ ★ ★

That may not be merely a gut feeling. One night in Shenzhen, I set up a Skype interview with Li Wang, the executive director of China Labor Watch. Li himself was a former Foxconn worker; he became a labor organizer and an advocate for better working conditions after living through the horrors at the company. He fled the country and now runs CLW out of New York City.

Li had high hopes for the chance at reform in the wake of the suicide epidemic and the resulting media spotlight. "Media reports are helpful," he says. "In 2011, when Foxconn abuse was reported by the media and could be asked about the suicide issue, wages rose almost one hundred percent and working conditions also improved. I think it's because of media pressure that Foxconn raised wages." In 2009, he says, the average worker's wages were around 1,000 renminbi ($145) a month, and by the end of 2010, it was raised to 2,000. "But after that, media transferred their attention to other subjects," he says. "Comparing 2013 and right now, nothing has been changed. Apple might have done a little bit in the beginning, but compared to what they promised, that's too little."

★ ★ ★

Back at Longhua, Wang and I set off for the recruitment center and the main worker entrance. Xu had called his friend Zhao, who still works at Foxconn—he had been promoted to floor manager years ago, and he had agreed to try to use his limited authority to get us past security for a tour.

He told us that he thought iPhones were made in factory block G2, in case we got in.

We wind around the perimeter, which stretches on and on—we have no idea this was barely a fraction of the factory at this point.

The factory walls loom over one side of a busy street; the other gives way to Shenzhen blocks and shops. A cheap LED billboard announces the recruitment center; it broadcasts images of cheerful workers at computer stations, quick shots of colorful assembly lines, footage of big blue swimming pools, large empty gyms, and nice-looking, clean buildings. It reeks of a Fox Trap.

Still, there are a handful of young men and women scribbling on forms as we go past, walk-in recruits like Xu once was. We turn left, past the center, and see another guarded entry station not far away. That is where we'd meet Zhao. But as we walk past the recruitment office, there is an entrance into a much larger, open space — no one is around, so we walk in.

★ ★ ★

The welcome center is an expansive, green-floored auditorium lined with eighty or so flat metal benches. A blue temporary wall cuts the space in half; it resembles a giant high-school gymnasium set up for a motivational speaker. Zhao later confirms that this was where Foxconn held the introductory presentation to workers. Beyond it lies a web of cubicle-size spaces, some with plastic test tubes and containers, probably where prospective factory workers undergo their mandatory health checkups before starting their jobs. Posters on the wall tout the reach and influence of Foxconn and the number of countries it has offices in. Another informs readers via cheery cartoon figures of policemen and hidden cameras that they are being surveilled.

The place was organized for mass-processing. Hundreds of recruits could be signed on at once here, dozens at a time given basic checkups or entrance interviews. We nose around until we reach a hall with two women behind the sort of plexiglass booths you see at the movie theater; they ask us what we were doing there, at which point we promptly leave.

Down the road, at the access point, we call Zhao; he says he'd be there in an hour. These security guards look a little nicer than the last set, so we ask if we can have a tour. They smile and said no. Any tour would have to be approved by the executive staff; they couldn't approve such a thing. We tell them we have a meeting with a floor manager, and they smile and repeat the same.

When Zhao shows up—a trim, nicely dressed man in his mid-twenties, with kind lines on his face—the story is the same. Executive approval required. Zhao had worked at Foxconn for eight years and had been a manager for a number of them, but that wasn't enough. No one goes in without the executive okay. There are too many secrets, the guards tell us. We could apply for approval online, although the process usually takes months. We spend nearly an hour trying to convince the security guards to let us in.

Eventually, we give up and walk the perimeter of the sprawling plant with Zhao, who has to get back to work in a different part of the facility. I ask him, as a veteran Foxconn employee, does he think it is really as bad as we've heard? Are the stories true?

"Everything you have heard is true," he says with a slight shake of his head. For a man whose job we had just heard required publicly humiliating his underlings, he seems far too kind and easygoing; there is nothing stern about his disposition, no itchy chip on the shoulder so many middle managers seem to tote.

"Then why work here?" I ask.

"I have adapted." He smiles and shrugs. "I do not scold my workers, like many managers do. I don't want to give them a hard time." He implies that his lenience might have prevented him from being promoted. I'm getting why Foxconn-hating Xu likes him. Zhao says he has just sort of settled into a career, though he doesn't seem thrilled about it. "Besides," he says, "I do not know what else I would do. I have been here so long."

After walking with Zhao along the perimeter for twenty minutes or so, we come to another entrance, another security checkpoint. There are apparently eight main ones and a handful of smaller ones. We say good-bye and watch him scan his card and disappear into the crowd.

That's when it hits me. I have to use the bathroom. Desperately. And that gives me an idea.

There's a bathroom in there, just a few hundred feet down a stairwell by the security point. I see the universal stick-man signage, and I gesture to it. This checkpoint is much smaller, much more informal—maybe an entryway for managers like Zhao? There's only one guard, a young man who looks something beyond bored.

Wang asks something a little pleadingly in Chinese. The guard slowly shakes his head no, looks at me. The strain on my face is very, very real. She asks again—he falters for a second, then another no.

We'll be right back, she insists, and now we're clearly making him uncomfortable. Mostly me. He doesn't want to deal with this.

Come right back, he says.

Of course, we don't.

Like I said, I can't believe it. To my knowledge, no American journalist had been inside a Foxconn plant without permission and a tour guide, without a carefully curated visit to pre-selected parts of the factory to demonstrate to the media how okay things really are. I duck into that bathroom, my head spinning. I can barely nod at the bewildered-looking kid washing his hands who's not even trying not to stare at me. I forget to go, slink out the door, and wave to Wang.

We power-walk through a factory block, then another, and another, and before I know it, we're at the end of the road, where a crumbling stone wall divides the factory grounds from the surrounding city. No one seems to be following us. Apartment highrises, a handful of trees, and a gray horizon complete the view. We hang a right alongside the wall, moving farther into the grounds. My adrenaline is surging; I have no idea where we are going.

Cinder blocks, gravel, and bricks are piled haphazardly around; a row of cones cordons off what looks like a spill. Blue trucks packed with shipping containers are parked here and there. Young men play a quiet pickup game of basketball in sweat-stained T-shirts. We move on, passing small streets that run inward and are lined with garages, workshops, and warehouse buildings. There's an official-looking building facing the yard with a stone gargoyle perched on either side of the door. I take out my iPhone and shoot some pictures of the

place where iPhones are made. The few people out here have started to stare.

We cut down one of the streets, past rusted, weather-streaked stalls. Some are filled with piles of raw materials, some stacked with cut metal, some held columns of empty pallets. A scratched-to-hell forklift sits wheel-less on blocks, emblazoned with graffiti. Once-white walls are a weather-beaten, erosive gray. It is, in other words, a lot like you'd imagine a shipping-and-receiving zone in any aging, city-size factory to be. A group of men on an elevated lift are drilling into the outside of a building, sending down showers of sparks. Half wear no safety gear. Debris spills out into the road, marked by a few red cones. Motorbikes and flatbeds dot the street.

As we make our way inward, the buildings get taller. Like a lot of cities, it gets denser the closer you get to downtown. Warehouses and workshops give way to two-, three-story buildings, then to what looked like dormitory high-rises. We start passing more people, each wearing an ID card on a necklace, who mostly side-eye us as we hustle on. The road widens to accommodate pedestrians and bicyclists, then cars too, and pretty soon, the way opens up into a busy intersection and a road crammed with hundreds, maybe thousands of young people. It looks like an exhibition or a jobs fair of some kind, but we don't stop to check it out. A couple of people stare at us, and a few hundred feet away, there is a security official directing traffic.

The gravity and the risk of the intrusion start to sink in. This is, clearly, a rash decision, as China isn't exactly known for its leniency toward journalists. There is no way we could ever hope to blend in, after all (there are no other lanky white Americans in sight). My translator, especially, could face harsh consequences if we are caught, but when I ask if we should turn back, she insists we push on.

We wait until the guard turns his back to address oncoming traffic and then walk past, trying to join the crowd.

Foxconn City really is a city.

We keep walking, and soon, the streets are lined with well-tamed shrubbery and shops and restaurants of every stripe. There are twenty-four-hour banks, a huge cafeteria, an open-air market that looks temporary but is mobbed with people. And there are people everywhere. Walking, riding, smoking, absorbed in their phones, eating noodles out of takeout boxes on the side of a road. Wearing polo shirts, jeans, plaid button-ups, stylish T-shirts, lanyards swinging around their necks, carrying their ID cards.

The streets are clean, the buildings newer here. Cartoon cat mascots give a thumbs-up over a storefront. Coca-Cola-branded umbrellas cover smartphone-browsing employees on metal picnic tables. Shiny sedans are parked in clearly designated parking spots along the main drag. There is a 7-Eleven—a fully branded, fully stocked 7-Eleven identical to every store in the franchise you've ever stepped foot in. For some reason, that bowls me over. We see what looks like cybercafés and strange inflatable structures designed to advertise the shops.

Together, it looks a bit like the university center of a college campus, just quieter. Given the sheer number of people, there is remarkably little noise. It's hard not to project after hearing horror stories all morning, but Longhua does seem infected by a ghostly, stifling air.

Maybe the most striking thing, beyond its size—it would take us nearly an hour to briskly walk across Longhua—is how radically different one end is from the other. It's like a gentrified city in that regard. On the "outskirts," let's call it, there's spilled chemicals, rusting facilities, and poorly overseen industrial labor. The closer you get to the "city center"—remember, this is a factory—the more the quality of life, or at least the amenities and the infrastructure, improves. In fact, one worker told us he did manual labor on the outskirts and believed he was paid less than the people who worked on consumer-electronics assembly lines.

As we get deeper in, surrounded by more and more people, it actually feels like we're getting noticed less. The barrage of stares mutate into disinterested glances. My working theory: The plant is so vast, security so tight, that if we are inside just walking around, we must have been allowed to do so. That, or nobody really gives a shit. We start trying to make our way to the G2 factory block, where Zhao had told us iPhones were made. After leaving "downtown" we begin seeing towering, monolithic factory blocks—C16, E7, and so on, many surrounded by crowds of workers.

This is when it starts to feel truly impressive. Look, a lot of factories skew dystopian; they are, after all, places constructed with the sole purpose of maximizing the efficiency of human and machine labor. But Longhua is different by virtue of its sheer expanse alone—it is block after block of looming, multiple-story, gray, grime-coated cubes. It is factories all the way down, a million consumer electronics being threaded together in identically drab monoliths. You feel tiny among them, like a brief spit of organic matter between aircraft carrier–size engines of industry. It's factories as far as you can see; there is simply nothing beautiful in sight.

In fact, the only things designed to be aesthetically pleasing, designed to appeal to humans at all, are the corporate mascots and the trimmed hedges back near the food court, and that feels grim out here—in Longhua, you're either in a strip mall or on the factory floor.

* * *

Foxconn City is a culmination of one of the very earliest human innovations—mass production. *Homo erectus,* which emerged 1.7 million years ago, were the first species to widely adopt tools and the first to become proficient at making them in large quantities. Some enterprising erectus hunters figured out how to make hand axes by rapidly striking several flint cores at once, in a feat Stephen L. Sass, a historian of materials science, calls "an early version of mass production."

It would take a few thousand centuries before that impulse would mature into the modern-day assembly line.

Imagine another factory. This one measures one and a half miles wide by one mile long, spans sixteen million square feet of factory floor space, and includes ninety-three towering buildings. It has its own dedicated power plant. It employs over a hundred thousand workers who toil for nearly twelve hours a day. Those workers have migrated from rural regions all across the country in search of higher wages. In all, it's a marvel of efficiency and production—it's described as an "almost self-sufficient and self-contained industrial city."

No, it's not run by Foxconn in the 2010s. It's Henry Ford's Rouge River complex in the 1930s. Even though Ford has been lionized as a hero of American industry, it's still easy to underappreciate the impact of the assembly line, an innovation perhaps more revolutionary than the iPhone or the Model T it now churns out at scale. And like most other innovations, it too had its bits that were borrowed from someone else, workshopped, tested, and sold to investors.

Ransom E. Olds (of Oldsmobile) had been running an assembly line for nearly a decade before Ford switched over to that mode of operation, though Ford's system contained numerous advances. Ford's biggest innovation, probably, was the supreme maximizing of efficiency. The distributed, station-based mode of production, in which each worker performs one specialized task ad infinitum, is what made complex machines like the automobile affordable and what makes the iPhone relatively affordable today. (It's also what gives Apple such large profit margins.)

But while we hold Ford and his mechanical assembly line up as a heroic example of American industriousness, it had roots in something much more organic — the slaughterhouse. The same Chicago slaughterhouses that incited national outrage after the publication of Upton Sinclair's *The Jungle* in 1906 were crucial to founding the operational system that produces the iPhone. Around that time, Ford's chief engineer, William "Pa" Klann, toured the Swift and Company slaughterhouse in Chicago. There, he saw what Ford would later refer to as "disassembly" lines, in which a butcher lopped the same cut of meat off each carcass that was passed down to him.

"If they can kill pigs and cows that way, we can build cars and build motors that way," Klann said. Ford engineers also toured the Westinghouse Foundry, which manufactured airbrakes and used "a conveyor system as early as 1890 to move molds into position," according to the historian David Hounshell. "We saw these conveyors in the Foundry and we thought, 'Well, why can't it work on our job?'" Klann recounted. The observation led to the now-infamous flow of production that would harness the power of repetition and machination, eventually allowing a Model T to roll off the line every twenty-four seconds by the 1930s.

And that, basically, is what's happening in China today, albeit with an even bigger labor force and an even more intricate, fine-tuned, and exhaustive labor operation. Consider this: Apple sold forty-eight million iPhones in the fourth quarter of 2015. Each and every one of those phones was assembled by hand, by a human being. Or, rather, by thousands of human beings. As of 2012, each iPhone required 141 steps and 24 labor hours to manufacture. It has likely risen since then. That means that, in a very conservative estimate, workers spent 1,152,000,000 hours screwing, gluing, soldering, and snapping iPhones together in a single three-month period. It's probably a lot more, given that large quantities of phones — sometimes as many as half — are scrapped because they don't meet quality standards.

In our interviews, the magic number we kept hearing was seventeen hundred—laborers charged with manning a machine stamp or checking the screens for quality said that's how many they were expected to oversee on a given workday, which averaged twelve hours. The same number came up for those tasked with cleaning them. Workers that were part of teams that tested the final phones said that, together, they were responsible for about three thousand phones a day. (Each earned around two thousand renminbi a month.) That adds up to more than two hundred iPhones per hour—over three a minute.

That is a herculean feat of manufacturing. Foxconn is now the world's biggest electronics-contracting company and the third-biggest technology company by revenue—its annual take is $131.8 billion—thanks largely to its iPhone orders. Specialized parts are still produced in other nations—processors come from the U.S., the chips and display panels come largely from Japan and Korea, the gyroscope comes from Italy, the batteries from Taiwan—but they're inevitably shipped to China to be assembled into an incredibly complex product-line Voltron.

And it's the ability to tackle that complexity with ruthless efficiency that makes Foxconn and its competitors so enticing to American companies like Apple.

In 2011, President Obama held a dinner meeting with some of Silicon Valley's top brass. Naturally, Steve Jobs was in attendance, and he was discussing overseas labor when Obama interrupted. He wanted to know what it would take to bring that work home. "Those jobs aren't coming back," Jobs famously said. It wasn't just that overseas labor was cheaper—which it was—it was also that the sheer size, industriousness, and flexibility of the workforce there was necessary to meet Apple's manufacturing needs.

In the *New York Times'* Pulitzer Prize–winning investigation into the so-called iEconomy, an unnamed Apple executive was quoted as saying that the real reason that Apple kept its operation overseas wasn't the cheap labor; some analysts estimated that building the phones in the U.S. would raise labor costs by only ten dollars a phone. No, they stayed there because of the immense, skilled workforce and the interlocking ecosystem of affiliated industry that had grown in Shenzhen. Droves of workers could be summoned to quickly assemble a new prototype for testing or swiftly make laborious adjustments to a huge number of products that were about to be shipped. Parts could be rapidly obtained and shepherded onto a production line. If Apple had to make a last-minute change to the iPhone—say, an alteration in the aluminum casing, or a new cut for the touchscreen—in a heartbeat, Foxconn could summon thousands of workers and hundreds of industrial engineers to oversee them.

The *New York Times* offered the following example:

> Apple executives say that going overseas, at this point, is their only option. One former executive described how the company relied upon a Chinese factory to revamp iPhone manufacturing just weeks before the device was due on shelves. Apple had redesigned the iPhone's screen at the last minute, forcing an assembly line overhaul. New screens began arriving at the plant near midnight.
>
> A foreman immediately roused 8,000 workers inside the company's dormitories, according to the executive. Each employee was given a biscuit and a cup of tea, guided to a workstation and within half an hour started a 12-hour shift fitting glass screens into beveled frames. Within 96 hours, the plant was producing over 10,000 iPhones a day.
>
> "The speed and flexibility is breathtaking," the executive said. "There's no American plant that can match that."

The follow-up question here might be: Why is it so imperative that our phones be assembled with "breathtaking" speed?

There are all sorts of MBA answers to that question—certainly, it gives Apple an operational advantage to be able to summon so many souls at the drop of a hat to mass-manufacture a new device or part. The rapidity of the process tightens shipments and allows Apple to be more nimble with matching production to demand—or effectively manipulating scarcity, even—and keeping extra inventory from piling up. It's cheap, efficient, and fast. It also aligns with Apple's instincts for secrecy: the less time a device spends in production, the fewer leaks there will be.

The dollar value of these advantages is considerable, but at the end of the day, the net difference between this massive, flexible operation and a more conventional assembly line that could be run in the U.S. amounts to a new phone getting into your hands a bit sooner and a bit more cheaply. The cost is tens of thousands of lives being made miserable by those last-minute orders, militaristic work environments, and relentless stretches of overtime. This is not necessarily Apple's fault, but it is certainly a by-product of a globalized workforce. Apple was actually one of the last major tech companies to move its manufacturing overseas; it had spent decades touting its Made in America bona fides.

And Tim Cook, who rose through the ranks at Apple on the strength of his supply-chain wizardry, is himself a key driver in that push toward breakneck production. One of his initiatives has been an attempt to eliminate inventory—today, Apple turns over its entire inventory every five days, meaning each iPhone goes from the factory line in China to a cargo jet to a consumer's hands in a single workweek.

Since the explosion in the iPhone's popularity—and the rise of the iPad and competitor smartphones and tablets—Foxconn has branched out and set up a number of factories across China. Longhua is likely still the biggest single factory operation, though today, a newer operation in Zhengzhou, a poorer, more rural region in mainland China, is the largest iPhone maker. According to a 2016 *New York Times* investigation, the Zhengzhou plant, now called "iPhone City" by locals, can churn out half a million handsets a day. Meanwhile, Foxconn is in talks with the Indian government to move some iPhone manufacturing to the second-most-populous nation; it already has factories running in farther-flung locations like the Czech Republic and Brazil, and it's considering more. It is reportedly building a fleet of so-called Foxbots, iPhone-building robots that might eventually replace human laborers altogether.

All of this serves to keep its employees' wages—which are higher than other factory workers', it must be noted—low. It's a pretty astonishing sign of how far the assembly line has evolved. Henry Ford famously began paying his workers five dollars a day in 1914, a high wage for the era, saying he thought they ought to be able to afford the Model Ts they were making. (That wasn't the whole story, of course—before he increased wages, he had a major attrition problem. The annual turnover rate was 370 percent because people hated the boring, repetitive work.)

That's not true of employees who make iPhones—despite the fact that it's only a handheld device, not a car. If an iPhone factory worker wants to buy the product he spends most of his waking life piecing together, he'd have to work several months straight—or find one on the black market.

Take the curious bazaar outside the Shanghai iPhone factory, marked by a banner reading PEGATRON MARKET. Yes, you can buy iPhones there. But they probably won't come from the megafactory next door. Most of them will come from around the world. One shop owner told us that he has an associate buy iPhones from the United States, where they can be purchased without an import tax, so he can sell them at lower prices.

Let that sink in for a minute. After all the myriad parts and materials flow to China from around the world—glass from Kentucky, sensors from Italy, chips from around China—after they are finally gathered together in one place, then pieced together bit by bit into the one device—here, in the Pegatron megafactory—the iPhones are sealed up and loaded onto a cargo plane bound for the United States. There, they are loaded onto the shelf of an Apple Store, where an enterprising Chinese associate buys them for U.S. prices and hauls them all the way back to Shanghai, literally a stone's throw from where they were manufactured. And that's the cheapest way to get a new iPhone for the workers who actually assemble them. I ask the seller what he thinks about the fact that the iPhone is made a couple hundred feet away, yet he has to buy them from America.

"I have no choice!" he says. "That's what I need for my business." And, indeed, his prices are nearly a hundred dollars cheaper than they are at a seller a few stalls down, who said he went through official channels to buy the iPhones from Apple.

This isn't a rare occurrence in Shanghai. We visited an ultramodern downtown mall, one that sold luxury goods, brand-name clothes, and upscale toys, where a number of stalls were advertising themselves as Apple Stores; they even had the white logo and the minimalist, light-wood table design. But a couple of the salesmen there openly admitted to us that they didn't get the phones from Apple, nor were they official Apple resellers. They too told us that they imported most of their iPhones from the U.S. — one said he had a network of college students living abroad who brought phones back to China for him — or turned to other means. One man told us that he had a contact inside Foxconn's Apple operation who supplied him with phones that "fell off" the trucks. This was just how things worked, they said — they clearly made little effort to hide their operations, based as they were inside a glamorous mall just a block or two from one of Shanghai's biggest metro stations. They even admitted that their shirts, which had AUTHORIZED APPLE RESELLER printed across them, were just for show.

"Everyone wants Apple, that's why we do it. I don't even like the iPhone," Xuao, who runs one such Apple Store, told us. He said he's not at all worried about Apple finding out. "For the Chinese, they tax the iPhones twice," he said. "First in Shenzhen when they make them, then at the border when they sell them. It makes no sense." A new 16 GB iPhone 6s can cost six thousand renminbi in

China—about a thousand dollars. Without operations like these, refurbished phones, or black-market phones, few in China's working class would be able to afford one.

"Everyone wants one," the assembly-line worker Jian tells me, "but there's no internal price for employees, so no one can afford one." Almost everyone we spoke with really liked the iPhone. They just couldn't afford it. In fact, whenever we asked if a worker had one, he or she would usually respond with a laugh and "Of course not."

Unlike in Ford's factories, Chinese assembly workers making ten to twenty dollars a day (in 2010s dollars) would have to pay the equivalent of three months' wages for the cheapest new iPhone. In reality, they'd have to scrimp and save for a year—remember, many workers barely make enough to live on unless they're pulling overtime—to be able to buy one. So none of them did. We didn't meet a single iPhone assembler who actually owned the product he or she made hundreds of each day.

★ ★ ★

There it is: G2. It's identical to the factory blocks that cluster around it and that threaten to fade into the background of the smoggy static sky. The crowds have been thinning out the farther away from the center we get; we've passed the entry point we tried to get through earlier, the road with the recruitment center on the other side of the factory walls. At this point, I've loosened up; we cruise past security guards, most of whom don't bother to look us in the eye. I worry about getting too cavalier and remind myself not to push it; we've been inside Foxconn for almost an hour now.

G2 looks deserted, though. A row of impossibly rusted lockers runs outside the building. No one's around. The door is open, so we go in. To the left, there's an entry to a massive, darkened space; we're heading for that when someone calls out. A floor manager has just come down the stairs, and he asks us what we're doing. My translator stammers something about meeting with Zhao, and the man looks confused — then he shows us the computer-monitoring system he uses to oversee production on the floor. There's no shift right now, he says, but this is how they watch. It looks a little old-fashioned; analog dials and even what looks like cathode-ray screen. It's hard to say; it's dark, not to mention damp in there, and my heart's racing again.

No sign of iPhones, though. We keep walking. Outside of G3, teetering stacks of black gadgets wrapped in plastic sit in front of what looks like another loading zone. A couple of workers on smartphones drift by us. We get close enough to see the gadgets through the plastic, and, nope, not iPhones either. They look like Apple TVs, minus the company logo. I should know — just the week before I left for China, I bought one. There are probably thousands stacked here, awaiting the next step in assembly line or waiting to be touched up and shipped out. We try the door, but this one's locked. We try a couple more — most end up being locked. Some are so rusted over, it's hard to imagine they can function as doors at all. Previous reports had stressed that workers, especially on Apple product lines, had to badge in before entering their factory floor; I wouldn't expect to be able to waltz in. Then again, we hadn't expected to stumble onto the grounds either.

But here we are, passing the hull of another building housing another operation piecing together another gadget. It's just so big. This isn't all Apple, of course; Foxconn helps manufacture Samsung phones, Sony PlayStations, and devices and computers of every type.

The infrastructure appears strained again, and while there's no construction or outdoor manual labor going on over here, the environs are definitely looking the worse for wear. If this is indeed where iPhones and Apple TVs are made, it's a fairly aggressively shitty place to spend long days unless you have a penchant for damp concrete and rust. The blocks keep coming, so we keep walking. Longhua starts to feel like the dull middle of a dystopian novel, where the dread sustains but the plot doesn't, or the later levels of a mediocre video game, where the shapes and structure start to feel uglily familiar, where you could nod off into a numb drift.

Soon, the buildings we reach begin to look downright abandoned. More lockers, cracked and rusted. Some teenagers wander past, clearly seeking out this periphery; they resemble the troop of kids in *Stand by Me*. We ask them where we are, and they shrug like teenagers.

"Here? They call this the docks," a girl says, and her group shuffles on.

They didn't necessarily look underage — an issue that Foxconn has grappled with in the past. In 2012, Foxconn admitted that up to 15 percent of its labor force during summer months were unpaid "interns" — 180,000 people, some as young as fourteen years old. While Foxconn insists that the work was purely voluntary and that students were free to leave, multiple independent reports revealed that vocational schools from around the region were forcing their students to man the assembly lines or drop out of school. Why the mandatory work assignment? To plug a labor shortage created by rising demand for the iPhone 5. After the reports surfaced, Foxconn vowed to reform its internship program, and, to be honest, I didn't see anyone that struck me as younger than sixteen.

We could keep going, but to our left, we see what looks like large housing complexes, probably the dormitories — complete with cage-like fences built out over the roof and the windows — and so we head in that direction. The closer we get to the dorms, the thicker the crowds get, and the more lanyards and black glasses and faded jeans and sneakers we see. College-age kids are gathered, smoking cigarettes, crowded around picnic tables, sitting on curbs. It's still quiet and subdued, like everyone's underwater. Hundreds of thousands of people and it never gets louder than the decibel of polite conversation.

And, yes, the body-catching nets are still there. Limp and sagging, they give the impression of tarps that have half blown off the things they're supposed to cover. I think of Xu, who said, "The nets are pointless. If somebody wants to commit suicide, they will do it."

We are drawing stares again — away from the factories and shops, maybe folks have more time and reason to indulge their curiosity. In any case, we've been inside Foxconn for an hour. I have no idea if the guard put out an alert when we didn't come back from the bathroom or if anyone's looking for us or what. The sense that it's probably best not to push it prevails, even though we haven't made it onto a working assembly line. Probably also for the best.

We head back the way we came. Before long, we find an exit. It's pushing evening as we join a river of thousands and, heads down, shuffle through the security checkpoint. Nobody says a word.

* * *

Getting out of the haunting megafactory is a relief, but the mood sticks. No, there were no child laborers with bleeding hands pleading at the windows. There were a number of things that would surely violate U.S. OSHA code — unprotected construction workers, open chemical spillage, decaying, rusted structures, and so on — but there are probably a lot of things at U.S. factories that would violate OSHA code too. Apple may well be right when it argues that these facilities are nicer than others out there. Foxconn was not our stereotypical conception of a sweatshop. But there was a different kind of ugliness. For whatever reason — the rules imposing silence on the factory floors, its pervasive reputation for tragedy, or the general feeling of unpleasantness the environment itself imparts — Longhua felt heavy, even oppressively subdued. Besides the restaurants and the cybercafés — both, notably, places where workers have to pay to hang out — there was no place designed in the interests of public well-being, or even designed to be an actual public space.

What was remarkable about Foxconn City was that the whole of its considerable expanse was unrepentantly dedicated to productivity and commerce. You were either working, paying, or shuffling grayly in between. Consumerism condensed into a potent microcosm. Eating, sleeping, working, passing time, all in Henry Ford's food court. In hindsight, it almost felt like those kids wandering out past the docks were staging a tiny resistance.

When I look back at the photos I snapped, I can't find one that has someone smiling in it. It does not seem like a surprise that people subjected to long hours, repetitive work, and harsh management might develop psychological issues. That unease is palpable; it's worked into the environment itself. As Xu said, "It's not a good place for human beings."

★ ★ ★

Since the suicide epidemic began, Apple has made some public efforts to hold its suppliers more accountable for workplace conditions. It began conducting supply-chain audits, releasing compliance reports, and instituting some worker-friendly policies to address more egregious violations. In 2012, Apple's audits uncovered 106 child laborers working in Chinese factories; Apple terminated contracts with one supplier, a circuit-board-component maker that employed seventy-four children under the age of sixteen, and forced the company to pay the costs of sending the children home. Apple became the first tech company to join the Fair Labor Association, a network of businesses that seek to promote worldwide labor laws in order to ensure better workplace conditions. Suicides have slowed, but not stopped. Workers are still logging too much overtime, but child labor has decreased. Wages seem to have stagnated, and turnover is still high.

China Labor Watch remains deeply unsatisfied and claims Apple's gestures have largely been made in the interest of public relations. "Apple joined the Fair Labor Association, which helped Apple a lot," Li Wang says. "It reduced the Foxconn pressure. The Fair Labor Association made a lot of promises to us and to the public, but as far as we can tell they are all lies. They did not achieve any of their promises."

There are no vacations in Longhua or Pegatron, that's for certain. But bright spots are emerging for China's workforce. Laborers are slowly becoming better organized, and wildcat strikes are becoming more common. A generation of poorly treated workers is apt to transfer its knowledge to the next, and as with protests against pollution, the predilection for popular resistance is growing. There are still few meaningful worker protections—so-called labor unions have long existed, but their leadership is appointed by the state, and their power is nil—but many workers have seen the power of collective action. Advocacy groups like CLM, SACOM, and the China Labor Bulletin have succeeded in pushing the issue of workers' rights into the public consciousness. Meanwhile, the bulging middle class is becoming less tolerant of poor conditions and labor abuses. Li says one improvement is that workers are now regularly getting their final paychecks when they leave the factories, whereas previously they often did not. But the quality of life for the workers—the ferocious pace, the semi-mandatory long hours—has remained the same for years.

"Nothing has changed," Li says. These precedents are doubly important because Apple and iPhone manufacturing contracts have such a massive influence on the industry—and on working conditions at large. "I had a meeting with Samsung executives and they said they would just follow Apple," Li says. "That's what they told us—they would do whatever Apple did."

★ ★ ★

In Shanghai, I met a charming Taiwanese couple who, after hearing I would be heading to Shenzhen, implored me to visit their company's factory, which manufactures iPhone accessories in the city's heart. They thought I would like to see their new technology, called Ash Cloud.

They were right. It was something.

The factory itself looked nicer than average—clean, modern, efficient. The operation was a standard assembly-line process, where workers manned stations, picked up pieces from the conveyor belt, did their part, then put the items back, where they moved on. About four hundred and fifty workers were employed here, I was told. At the moment, they were producing nice iPhone cases for European markets like Italy.

But throughout the factory, vertical LED screens were hung between the stations. Each broadcast a worker's portrait in the upper left-hand corner beside a readout of numbers, then changed to a screen full of stats in a clean, iOS-friendly UI. It was, of course, part of an iPhone app. Using the Ash Cloud, executives or floor managers could track worker productivity down to the number of units produced, and they could do so remotely or from different parts of the floor.

If a worker's production rate slowed below the standard, the numbers turned red. If it was on or ahead of target, they turned green. And each time the worker successfully performed a task on an item passing down the assembly line, a number ticked up toward the quota.

They had done it. They'd closed the loop. They'd made an app for driving the workers that make the devices that enabled apps. They were hoping to spread the word, that licensing the Ash Cloud app could become another part of their business; a couple factories had already been using it, they said. Now, factory workers could be controlled, literally, by the devices they were manufacturing.

I thought of one ex-Foxconn worker we interviewed. "It never stops," he said. "It's just phones and phones and phones."

CHAPTER 13

Sellphone

How Apple markets, mythologizes, and moves the iPhone

The Bill Graham Civic Auditorium in downtown San Francisco, which can seat six thousand people, is going to be absolutely packed. I join the shuffling masses—tech journalists, Apple employees, industry analysts—and inch forward with the plaid-shirt-and-jeans-patterned glacier. It feels more like the entrance line to a rock concert. The lights are low, and people are genuinely excited. And so am I.

We're here because the iPhone 7 is about to be announced. This is all an elaborate, well-choreographed sales pitch, but I can't help being excited. News vans are parked outside, video cameras are angled to capture reporters with the giant Apple logo installed atop the auditorium over their shoulders, idle chatter buzzes, laptops are everywhere.

Product launches are a pillar of Apple's mythology/marketing machine. Steve Jobs introduced every major Apple product since the Mac from a stage like this one. When Aaron Sorkin wrote a film about Jobs, he set it entirely backstage at three product launches. The keynote speeches at the events became so closely associated with Jobs that fans took to calling them Stevenotes.

For good reason: Jobs was a master salesman. He didn't typically get up on the stage and tick off product specs or descend into the effusive marketing-speak his competitors and successors sometimes do. He wasn't telling you why you should buy an Apple product; he was matter-of-factly discussing the attributes of this Apple product that was about to change the world. His declarations felt natural, emphatic, and true. When he told you Apple was "revolutionizing the phone," he believed it. The tradition has persisted since his passing in 2012; Tim Cook has dutifully taken over the presenter-in-chief slot, though he clearly relishes it a little less than his predecessor.

This time the buzz isn't about what the next great addition to the iPhone will be—in the past, it'd been features like a front-facing camera, Siri, or a larger screen—it's largely about a big subtraction. For months, Apple blogs and tech sites had speculated that Apple was going to pull the plug on the headphone jack in an effort to anoint wireless headphones as the new norm.

I sit down next to Mark Spoonauer, the editor in chief of a trusted gadget-review site called Tom's Guide. He says he's been to at least seven Apple product launches, and he attends the Events to try to discern what really is new and "what's worth caring about," and to answer the ur-question for gadget blogs: Is it worth upgrading?

"Even if someone has done a feature before, Apple needs to prove that they can do it better. It's also about proving that Apple can still innovate in a post-Jobs world," he says. After years of attending these product-launch events, Spoonauer is still glad to get the email invite from Apple (the Event is invitation only). "There's still excitement about being here," he says. "It's not just about the product; it's about the atmosphere."

The lights go down, and a video rolls. It shows Tim Cook calling a Lyft for a ride to the Apple Event—the very event we are waiting for him to show up at—only to find that the car is being driven by James Cordon of Carpool Karaoke, who is then joined by Usher for some reason. They all sing "Sweet Home Alabama" together, and the flesh-and-blood Cook runs out onstage.

He makes some announcements, and then invites Shigeru Miyamoto, the legendary founder of Nintendo, up to the stage to announce the company's first foray into iPhone games, Mario Run. The crowd's rapt quiet gives way to enthused pandemonium.

Eventually, he gets to the iPhone. "It's a cultural phenomenon, touching lives of people all around the world," Cook says as the video feed cuts to a pan of the audience, which, of course, consists of hundreds of people staring at their iPhones. "It is the bestselling product of its kind in the history of the world."

Presentations like this—especially when they were given by Steve Jobs—are one of the major reasons that everything Cook is saying right now is true. Simply put, the iPhone would not be what it is today were it not for Apple's extraordinary marketing and retail strategies. It is in a league of its own in creating want, fostering demand, and broadcasting technologic cool. By the time the iPhone was actually announced in 2007, speculation and rumor over the device had reached a fever pitch, generating a hype that few to no marketing departments are capable of ginning up.

I see at least three key forces at work. Together, they go a little something like this:

1. *Shroud products in electric secrecy leading up to...*
2. *Sublime product launches featuring said products that are soon to appear in...*
3. *Immaculately designed Apple Stores.*

Of course, for any of it to work, the product itself has to be impressive. But creating a mythology around that product is, especially in the early stages, as important to selling it as anything else.

Traditional marketing campaigns are important too, of course, and Apple has run plenty of iPhone ads. There hasn't been a truly classic iPhone spot or campaign, on the level of the famous Ridley Scott–directed "Big Brother" ad that introduced the Macintosh during the 1984 Super Bowl, the "Think Different" ads that reminded audiences that the Apple brand was associated with geniuses and world-changers in the late 1990s, the earbuds-and-silhouette campaign that created an efficient aesthetic shorthand for iPod cool in the early 2000s, or even the "I'm a Mac," "I'm a PC" ads that played off Windows-based computers' reputation for being buggy and lame.

The closest thing the iPhone has to a classic is probably the "There's an App for that" campaign in 2008. The debut ad for the iPhone, "Hello," was a mashup of famous faces answering the phone and is largely forgotten today. Other early ads were largely explanatory, which makes them interesting to watch; they're artifacts from a time when the concept of browsing the web with your finger and then taking a call needed an introduction. One nicely executed and entirely prescient spot, "Calamari," shows a user watching a *Pirates of the Caribbean* clip of a giant squid attack, getting a hankering for seafood, switching to Google Maps to search for a place nearby, and calling the restaurant, all with a few finger taps. A sequence of actions like that was pretty revolutionary at the time. Others highlight the ease of surfing the "real" internet, listening to music, and using Facebook on the go.

Still, the majority of major corporations can afford well-produced ad campaigns, and even the most uncool can score the odd hit. In the absence of a definitive iPhone ad campaign, it's worth looking at what Apple does differently than its competition to elevate its marquee product.

So, number one: You can't talk about the iPhone without talking about Apple's secrecy. The way that Apple has honed its ultrasecretive approach to cater to and exploit the online hype machine is an innovation unto itself, one that rivals many of its other more tangible technological innovations. It too is steeped in history.

* * *

Apple is one of the most secretive companies in the world, and the imperative originated at the top. Jobs was always proactive in managing his company's media appearances; from the early days, he was keen on developing relationships with editors and writers at the major magazines and newspapers. But he wasn't always super-secretive. The *New York Times* reporter John Markoff, one of the writers who'd earned access to Apple, noticed the change in the late 1990s and early 2000s.

"Since Mr. Jobs returned to Apple, he has increasingly insisted that the company speak with just the voices of top executives," Markoff noted after being denied an interview with a driving force behind the iPod, Tony Fadell. Another *Times* writer, Nick Bilton, observed that Jobs frequently described his products as "magical," and, "as Mr. Jobs knew so well, one thing that makes magic so, well, magical, is that you don't know how it works. It's also one reason Apple is so annoyingly tight-lipped."

Harnessing secrecy to generate interest in a new technology isn't a novel idea. It's been a key element in ginning up interest in new commercial technologies—even the ones that seem in hindsight like obvious breakthroughs.

"Flight represented the pinnacle of human achievement," writes the technology historian David Nye. "To raise a heavier-than-air vehicle into the sky was a technological marvel, the fulfillment of a centuries-old dream. Yet when the Wright Brothers flew for the first time, in 1903, almost no one saw their achievement." They hadn't fed anyone's sense of intrigue; there was no anticipation. So the brothers changed tack. "The Wrights remained secretive about their plane's design during subsequent development, seldom allowing the press to see what they were doing," Nye explains. Word trickled out, and the Wrights let it. When they were invited to discuss their invention at the St. Louis World's Fair in 1904, they refused. "They had their eyes on commercial applications, and they were unwilling to disclose the details of their machine." The Wrights waited until 1908 and held a grand demonstration for the U.S. Army. "Huge throngs turned out to see them…Until the time of World War I, many people ran out of their houses to stare at any airplane that flew into view."

Similarly, clamping down on Apple's public doings was a conscious decision.

In *Becoming Steve Jobs*, Brent Schlender and Rick Tetzeli explain that Jobs "directed Katie Cotton, his communications chief at Apple, to adopt a policy in which Steve made himself available only to a few print outlets . . . Whenever he had a product to hawk, he and Cotton would decide which of this handful of trusted outlets would get the story. And Steve would tell it, alone." And he, of course, would keep the details close to his chest. Schlender, who covered Jobs for years, talked with him "many times about his reluctance to share the spotlight with the others on his team, since I asked repeatedly to speak with them and was largely unsuccessful." Jobs would say he didn't want his competitors to find out who was doing the best work for fear of losing them, which struck Schlender as "disingenuous." What he did buy was that "Steve didn't think anyone else could tell the story of his product, or his company, as well as he could."

The effect was to create a vacuum of official Apple news. As the company reemerged from its 1990s slump with a bevy of popular, sexy electronics like the sleek Bondi Blue iMac and the iPod, the demand for intel on the company's doings boomed. Fan blogs, industry analysts, and tech reporters all commenced circling the reawakening tech giant, turning Apple-watching into a full-time job.

Speculating on the rumored iPhone became a cottage industry unto itself in the mid-aughts, and the practice continues. "Apple is so secretive that there is essentially an entire industry built around creating, spreading and debunking rumors about the company," the *Huffington Post* declared in 2012. Indeed, there are too many Apple-dedicated blogs and websites to count; *Apple Insider, iMore, MacRumors, iLounge, 9to5Mac, Cult of Mac, Daring Fireball, Macworld, iDownloadBlog,* and *iPhoneLife* are a few. All of these publications, in serving their audiences, are both meeting a real demand in an iPhone-heavy world and giving the iPhone reams of free press.

* * *

So here's the thing about that annoying secrecy: It works. At least, it has for Apple, helping to elevate the status of the iPhone as a product apart. One former Apple executive estimated that keeping the first iPhone secret "was worth hundreds of millions of dollars."

How's that, exactly? In addition to the free press generated by Apple-dedicated sites, secrecy plays a powerful role in ratcheting up demand among consumers. In a 2013 paper for *Business Horizons*, "Marketing Value and the Denial of Availability," David Hannah and two fellow business professors at Simon Fraser University theorized how Apple's secrecy benefits its product sales. "According to reactance theory, whenever free choice — for example, of goods or services — is limited or restricted, the need to retain freedoms makes humans desire them significantly more than previously, especially if marketers can convince people that those freedoms are important. Apple applies the principle very effectively." Not only are product specifications and launch dates closely guarded secrets, the authors wrote, "the company also keeps supplies immediately post-launch artificially low." You can't know about it before it's launched, and you still can't get your hands on it once it's available.

So die-hard Apple fans turn to the live-stream or their Twitter feed to see the secrets of the new iPhone. That sense of revelation propels a sense of desire, which Apple exploits by introducing the new iPhone with highly regulated scarcity. Fans will "happily wait in line — often throughout the night — for stores to open in order to be among the first to purchase the new product, despite the obvious fact that it will be readily available in just a few weeks more."

That spectacle of lines of die-hard Apple fans stretching around city blocks of course further feeds the story about how in demand the iPhone is, which further feeds the gratification of all those who participate in the ritual of obtaining one.

After the first iPhone began its ascent to most-profitable product, period, secrecy inside Apple naturally only increased in its wake. Employees who leaked details about upcoming products could be fired on the spot. Teams charged with a project Jobs deemed especially important would be made to operate in secrecy, even among their peers.

Of all the complaints about working at Apple I gathered in the course of talking to the iPhone's architects, its secrecy was at the top of the list—engineers and designers found it set up unnecessary divisions between employees who might otherwise have collaborated.

Abroad, Jobs is said to have distributed false product schematics to Apple's suppliers in an effort to locate leakers—if the fake product showed up on fan site, Jobs would know the source of the leak and fire the supplier.

Tony Fadell, a senior vice president and once one of the company's stars, told me that at times, the secrecy made working on the iPhone—which he was in charge of hardware for—next to impossible.

"I saw the falling-out because of that, especially when it's such an incredibly hard program, and we all had to be working together, but yet, we weren't on the most critical pieces," he says.

That impulse to secrecy was transmitted to Steve's peers. "It was fueled by not just Steve but others who had the power that Steve gave them, and they wanted to make sure they secured it at all times, and they would not necessarily tell us stuff. They would make us intentionally look bad, and point to us, and we couldn't defend ourselves because we had no information."

Today, the company is much larger, and since Jobs's passing, it's under the command of a CEO with a less paranoid style. As Apple's supply chain has continued to expand, more leaks have dripped out, and Cook has shown less interest in punishing the leakers. "The leakers have gotten so much better over the years, there's not much left in the way of mystery," Spoonauer says, so when he comes to an Apple Event, "I'm more interested in how they're going to spin those leaks." So one might think that secrecy inside the company would fade as well. Apparently not.

"It's worse than ever," Brian Huppi, the input engineer who helped conceive the original iPhone design paradigm, tells me. He went back to Apple after a few years' hiatus and found that interdepartmental secrecy had reached new heights, before he left again.

Even current Apple employees at just about every level of seniority chafed at its near-total nondisclosure policies when I managed to get through to them. Many that I reached out to told me that they'd love to be able to sit for an interview, would love to discuss their contributions publicly, but that, per company policy, they couldn't talk.

I did meet with a representative of iPhone PR at Apple's HQ in Cupertino. "The only reason we're even talking to you," he said as we sat outside the cafeteria at a table in the middle of Infinite Loop, was that Apple was in the process of opening up. But it never really did.

★ ★ ★

One result of all that secrecy is that it allows Apple to more tightly control its message and keep the focus expressly on the products and away from its more controversial practices—conditions in the factories that manufacture the phones, say, or its offshoring of $240 billion in tax havens in Ireland. Or even less controversial things, like the role a particular employee played in developing the iPhone.

Apple has essentially cultivated a new set of norms among the public and the tech press—no access, no official comment, no transparency. So I called up the editor of the *Atlantic*'s tech section, Adrienne LaFrance, who had recently written a treatise on the neutering of the tech press. I wanted to hear how this Apple-led trend was affecting the public sphere.

"There gets to be this danger of when people expect the tech companies to not give you on-record interviews or not to ever comment, then you slowly get to this place where it's not clear to me always that journalists are really doggedly going after information—assuming, often correctly, that they're not going to get it," she says. So by denying journalists access for so long, Apple (and other increasingly secretive tech companies) trained them to accept the official line or the details doled out at the public-facing launch events.

"Everyone on all sides is getting too comfortable with this arrangement," she says. "If you look at the ecosystem of tech coverage, how much is dedicated to the evaluation of a product versus the practices of the company?" she says. It has positioned the product as the center of its universe. It exists almost apart from the world of workers, of developers, of users, of business.

So how do you crack the code? "Even if the answer is no every single time, you have to keep trying," she says.

So I did.

The *Register*, a UK-based tech pub known for its strident, critical views of the industry, ran a funny story detailing its employees' efforts to obtain an invite to the iPhone 7 Apple Event. They installed an email tracker to see if Apple's press folks were in fact reading their entreaties. It turned out they were. (They didn't get in.)

So I decided to do the same. Apple hadn't responded to my latest futile request for interviews for months. So, I installed an email tracker made by a company called Streak, and I sent a fresh query. By the end of the day, it had been read on three different devices, presumably by three different people. I never heard back. I tried again a week later, with the same result. Nice.

Eventually, I decided to cut out the middle man and write directly to Tim Cook. You never know, right? Jobs was famous for randomly responding to notes in his in-box, and Cook had done the same once or twice.

I sent Tim Cook an email requesting an interview on August 31, 2016. That's when things got interesting. The tracking software I installed works by loading a tiny, transparent 1x1 pixel into an email message. When it's opened, the image pings the server it came from with data that includes the time and location that the email was opened as well as the kind of device used to open it.

That was the weird thing. When Tim Cook opened my email, the software showed me what kind of device he'd opened it on: A Windows desktop computer.

That couldn't be right. I emailed Streak to ask how accurate that part of the service was. Their support team told me, "If it has specific device data: Very accurate." I sent a follow-up email to Cook. Once again, it was opened — on a Windows desktop computer.

Was Tim Cook using a PC? Or was whoever was sorting through his emails? Either possibility seemed odd.

Apparently, the email I sent to Cook made its way to Apple PR; my

Hail Mary had been hailed. I asked the PR rep if Tim Cook had actually opened my email. "Yes," she said, "he read it and forwarded it on."

Okay, then. A couple weeks later, I sent one more follow-up. It was opened, again, on a Windows desktop computer. He never did write back.

★ ★ ★

Okay, okay. So we have a company that has long put an emphasis on extreme secrecy, giving rise to media that feverishly reports on anything and everything Apple-related, giving rise to a core of user-consumers awaiting the latest release. Sounds like the groundwork has been laid for a well-honed message—the entry of a definitive voice that can correct the record once and for all and excite the masses anew.

And so we get the biggest public displays that Apple offers—the invitation-only, tech-demo spectacle of the storied Apple Event.

These Stevenotes aren't a novel format. Alexander Graham Bell, recall, went on tours and put on shows in exhibition halls and convention centers across the Eastern Seaboard to demonstrate his new telephone.

But the most famous tech demo of all was the one that may have most informed Jobs's style.

In 1968, an idealistic computer scientist named Doug Engelbart brought together hundreds of interested industry onlookers at the San Francisco Civic Center—the same civic center where the iPhone 7 demo was made nearly forty years later—and introduced a handful of technologies that would form the foundational DNA of modern personal computing.

Not only did Engelbart show off publicly a number of inventions like the mouse, keypads, keyboards, word processors, hypertext, videoconferencing, and windows, he showed them off by using them in real time.

The tech journalist Steven Levy would call it "the mother of all demos," and the name stuck.

A video feed shared the programs and technologies being demoed onscreen. It was a far cry from the more polished product launches Jobs would become famous for decades later; Engelbart broadcast his own head in the frame as, over the course of an hour and a half, he displayed new feats of computing and made delightfully odd quips and self-interruptions.

"As windows open and shut, and their contents reshuffled, the audience stared into the maw of cyberspace," Levy writes. "Engelbart, with a no-hands mike, talked them through, a calming voice from Mission Control as the truly final frontier whizzed before their eyes. Not only was the future explained, it was there." The model for today's tech-industry keynote presentations was forged, almost instantly; the presentation style was perhaps not as influential as the technologies presented, but they were closely intertwined.

Through his suite of inventions, which were further developed at Xerox PARC—yes, PARC again—Engelbart laid the foundation of modern computing. But he insisted that PCs were antisocial and counterintuitive; his dream was augmenting the human intellect through collaboration. He imagined people logging on to the same system to share information to improve their understanding of the world and its increasingly complex problems. He advocated something a lot like the modern internet, social networking, and a mode of computing that, through the smartphone, has indeed begun the supplanting of the PC as the primary way we most often trade information.

Though Engelbart's mother of all demos became legendary among the computer crowd, it was an outsider, it seems, who would turn Steve on to the format he later became famous for. Apple expert Leander Kahney says that Jobs's keynotes were the product of CEO John Sculley: "A marketing expert, he envisioned the product announcements as 'news theater,' a show put on for the press. The idea was to stage an event that the media would treat as news, generating headlines for whatever product was introduced. News stories, of course, are the most valuable advertising there is." Sculley thought that entertaining a crowd should be the priority, so product demos should be "like staging a performance," he wrote in his autobiography, *Odyssey*. "The way to motivate people is to get them interested in your product, to entertain them, and to turn your product into an incredibly important event."

Combining exciting new technologies with theater has become a uniquely American art form, and Apple has perfected it. It taps directly into what the historian David Nye calls "the American technological sublime"—the awe people feel at witnessing an impressive new leap in technology. Although America is a diverse nation, fragmented in religious belief and cultural values, its citizens have long found common ground, Nye argues, in the uniting power of an impressive new technological feat. The Hoover Dam, the lightbulb, the atomic bomb. We find solidarity in the language of areligious, asexual progress.

And it works. You feel it at the civic center as tech executives walk onstage brandishing the latest world-changing gadget. And the secrecy generated beforehand, the sense that you're being allowed a peek under the hood is—undeniably—a little bit thrilling.

But as gadget-review editor Mark Spoonauer reminds me, "There are journalists who actually try to stay away from the 'reality distortion field,' because what you don't want to do is get caught up in the excitement. Because you have to be objective."

It's just really hard to do after Apple delivers you the sublime.

* * *

After the presentation, which concluded with a performance by the Australian pop singer Sia, who stood motionless in a giant wig and sang her hits while a kid bounced around and did cartwheels, the press is funneled into a room, stage right, that resembles a miniature version of an Apple Store—a month into the future, when the products just announced onstage will be available. We all get our first shot at handling, swiping, and snapping photos with the iPhone 7. I tried on the new AirPods—the new wireless earbuds, which, in my head, I could not refrain from thinking of as Airbuds—and piped in some Apple-sanctioned tunes.

Bloggers and news crews were angled everywhere, filming stand-ups in front of the products, rattling off first reactions. Others were jotting down notes. There must have been a hundred blog posts filed from the premises that hour. More people kept cycling in, and the room seemed increasingly crowded. There were lots of people taking photos of the iPhones with their iPhones, and people like me, using their iPhones to take photos of the people taking photos of iPhones with their iPhones.

It was a curious simulacrum; an Apple Store turned into a showroom, a showroom of a showroom. The space designed into synonymity with modern retail done up as a celebrity. This is what these products would look like out in the wild. And so they would.

* * *

Weeks later, on the day the iPhone 7 was slated to launch at retail stores around the nation, I set out to see the results of that marketing machinery in action. I made my pilgrimage to one of the first Apple Stores. This location in the Glendale Galleria outside of Los Angeles, along with the store in Tysons Corner Center in Virginia, were the first to open, on May 19, 2001. I was meeting a friend, Jona Bechtolt, a die-hard Apple fan—he's even got the Apple logo tattooed on his leg—who was planning on upgrading to a 7 that day. I wanted to see if crowds still turned out in droves, if those famous lines would stretch on, nearly ten years after the 2007 iPhone inspired the first queues to became media sensations.

Short answer: Yep.

The line stretched out across the entryway, through the central corridor, and around the corner here on the second story of this indoor mall. It certainly wasn't the size of the epic, block-long lines of yore; I counted forty people. Still, that was a lot for an iPhone model that a lot of the press had written off as a nonessential upgrade.

As soon as I walked up, I heard the sound of three-quarter-hearted, corporate-colored cheering. The doors had just opened, and, as is customary for Apple Stores on launch day, the employees line up and applaud the customers who were dedicated enough to show up hours early, or even spend the night. A handful had.

"I do it every year," a man named John said with a smile, and almost a shrug. The crowd was a mix of die-hard enthusiasts who still enjoyed the ritual of waiting outside overnight to be among the first to own the latest iPhone even if it wasn't necessary, and mini-entrepreneurs who planned on buying the maximum allotted number and reselling them to friends and on eBay while supply was still constrained and the phones were in high demand, hoping, as in years past, that phones would sell out for a couple weeks. "I'm gonna buy eight, sell two to my friends, and do eBay for the rest," one woman told me. "I take a little on top."

This year, the jet-black phone (a new color) and the 7 Plus (the larger model with a new dual camera) both sold out early. "Selling out" is part of the dance. The first iPhone was "sold out" for the first couple months after it went on sale in 2007, and we'll never know if that was due to a legitimate supply shortage. That does seem plausible, given the rush involved in getting it out. But for later models, Apple's finely tuned supply-chain and its sway over suppliers means that most scarcity in subsequent launches is plausibly artificially generated by Apple.

"It's not just design, it's just not the iPhone, it's not just the marketing," Bill Buxton says. It's also about maintaining supply-chain flexibility and inflating demand by creating the impression of scarcity. It's about making the iPhone feel "like the Cabbage Patch doll, everybody is running out to get one and buying them because they were afraid that they were going to be out of stock and they needed to give one to somebody for Christmas. I do not know a single person, I challenge you to find a single person, who despite that feeding frenzy could not find one." Which is a good point. We know now that Apple's suppliers can manufacture half a million phones in a day and ship them to the U.S. in another one. Believing that the most anticipated new color of the new model has sold out requires a suspension of disbelief.

"It's completely manufactured by one of the most brilliant marketing teams," Buxton says. "They designed the production, supply chain, and everything. So with the greatest tradition of Spinal Tap, if needed, they could turn the volume up to eleven to meet demand."

Anyway, I was surprised—in 2016, I hadn't expected to find anyone camped out or willing to wait in line for hours, given the tenor of the conversation around the iPhone 7, which hadn't generated as much excitement as previous models. But there they were, spending the better part of a day outside an Apple Store.

There are worse places to be. Apple's immaculately designed product hubs are the envy of retail stores around the world. Intended, it is said, to resemble the long wooden tables used in Jony Ive's Industrial Design Lab, with considerable input from Jobs, who holds a patent on the glass staircases, Apple Stores began opening in the early aughts. Initially opposed by the board, they've proven to be a sales behemoth.

In 2015, they were the most profitable per square foot of any retail operation in the nation by a massive margin; the stores pulled in $5,546 per square foot. With two-thirds of all Apple revenues generated by iPhones, that's a lot of hocked handsets.

And the Geniuses and Specialists doing the hocking make up a large number Apple's employees. Across 265 U.S. stores, Apple says it has thirty thousand retail employees. As of 2015, that was nearly half of the company's total U.S. workforce. Given the high volume of sales and the immense success of the retail spaces, these are some of the most productive retail workers in the nation.

In 2011, the Apple analyst Horace Dediu broke down the numbers in an attempt to calculate just how productive. He found that, on average, each employee at a U.S. Apple Store generated $481,000 in 2010 and was on track to do roughly the same in 2011. That's nearly four times as much as employees made for JC Penney, he noted. Average employees served six customers an hour and generated about $278 per hour.

Apple retail employees made from nine to fifteen dollars an hour and received no commission for those sales. While that's well above the minimum wage, the company's skyward profits put the relatively low wages into stark contrast. Apple had no trouble attracting employees; the iPod and then the iPhone had made Apple popular among precisely the young-skewing set that was ideal for the company, enough so that it would attract criticism for resembling a cult. But it was developing a retention problem, due in part to the low-end, commission-free wages.

As a 2012 *New York Times* headline put it, "Apple's Retail Army, Long on Loyalty but Short on Pay." Health benefits were available only to full-time employees, and the advancement structure was arcane. Tensions began to rise inside the perpetually optimistic-beaming company.

The retail stores were designed to be beautifully stark tech sanctums, places that would inspire a little awe in consumers and cast Apple products as the tools of the future. And the enthusiastic Geniuses and Specialists were tireless Apple ambassadors, instrumental in extending its message to consumers, in creating an environment where consumers would be thrilled to participate in that future and buy new iPhones. They also had to educate new Apple gadget owners, diagnose problems with existing products, fix them if possible, and tend to the more mundane demands of retail. It's hard work, in other words. And behind the scenes, there is a human cost to the carefully constructed retail ritual.

One employee, a part-time Specialist at Apple's flagship San Francisco store, decided to make a stand.

* * *

"As much as we helped Apple to be a really cool place to come in and shop for things, we wanted it to be a fun and enjoyable place to work," Cory Moll tells me. "And it was becoming less of that." Moll had been working for Apple since 2007; he started at the Madison store in Wisconsin, where he was from. In 2010, he transferred to the flagship Apple Store in downtown San Francisco. He's a die-hard Apple fan; he tells me he can't wait to get the iPhone 7 and is considering jet-black but was worried it'd scuff up.

And he sighs, just thinking aloud about the incoming rush (we spoke the week before the 7 was set to hit stores). "That's going to be a whole lot of crazy, happy fun. I miss being a part of that," he says. "Launch days—iPhones are always the biggest. Any Mac updates, people come in for that. But the iPhone was where it's at."

But after being at Apple for a few years, and working at its flagship San Francisco store, he began to see some systemic problems.

"Pay was an issue," he says. "Compared to other companies in other regions, for as long as we'd been there, only seeing raises of one, two, three percent, that's a small number." And it didn't seem to reflect the employees' expertise and skill set, the familiarity with the products, with Apple culture, the salesmanship. "We all had developed a strong skill set and knowledge base," he says, so "making twelve dollars an hour on top of not having any benefits, that's kind of saying, 'Hey, you're working for one of the top companies in the world, and you're barely making minimum wage, and if you get sick, well, screw you.'"

There was no mechanism to discuss promotions, and management would schedule part-time workers to do full-time weeks without offering them the status change that would let them qualify for benefits. When Moll or his co-workers asked about the longer hours, management would simply cut them. Moll alleges that practices like that could be violating labor law, by misclassifying workers.

"In terms of scheduling, in terms of promotion—people who had been there for years and years and years—being overlooked for full-time status, being overlooked for role changes. From being a Specialist to being a Genius. It was incredibly difficult. It felt like there was a lot of favoritism, when it comes to being looked at for a promotion, you've got to be friends with management team, you've got to be buddy-buddy."

Many of Moll's peers felt the same way. "Working stressful hours, you don't really know two weeks out from next when you're going to be working. And then if there's a launch event, that throws everything off." After Ron Johnson, whom many retail employees saw as the father of Apple retail, left to join JC Penney in 2011 and was replaced by Jon Browett, who brought a colder, outsider style, those tensions started to boil over.

"Not a single person I knew liked the direction he was taking it," Moll says. He began to grow interested in the idea of organizing, though he wasn't sure how to do it yet. But he started discussing the possibility with his peers.

"The conversations I had with people on the inside, they varied, of course," Moll recalls. "Some people were excited about it, and there were people who were afraid of it. I didn't really position it as wanting an official union. I really positioned it as us getting together, and whatever that looked like—we could figure that bit later on. I was really just focused on building a voice."

It quickly became apparent that it would be difficult to discuss, much less organize, around the bustling Apple Store, so Moll turned to Twitter—he started reaching out to employees and organizing support over the social network. After he figured he'd received a show of solidarity from "a couple hundred" local and national Apple retail workers, he crafted a press release and blasted it out to the tech press. He set up a website, AppleRetailUnion.com, and met with established unions that were interested in helping them organize. Eventually Moll made himself known as the driving force behind the effort and did interviews with the likes of CNET, the *Times*, and Reuters. And he set up an electronic form that would allow Apple retail workers to submit grievances to him, and he would forward them on to corporate—he got hundreds of complaints that way.

Apple, of course, responded in kind. The company disseminated "union training materials" to its stores, which were largely interpreted as tools to help management quell union activity. Shortly after, however, it announced that it would be awarding raises early, offering more training opportunities to part-time staff, and extending benefit packages to part-time workers.

Those raises did in fact materialize. Moll says his pay was bumped up by $2.42 an hour, a much bigger increase than usual, and most workers saw increases of that size too. It was, undeniably, a victory for the thousands of iPhone sellers who helped Apple turn immense profits. "I know going public with what we want to have happen definitely lit a fire under their butts and said, 'Hey, we really need to reconsider how valuable these people are to this company.'"

It also, however, deflated interest in the unionization drive. After five and a half years at Apple, Moll decided it was time for a change, and he left the company.

Even though Moll's drive didn't result in a recognized union, the effort did improve the quality of life for thousands then and to come. "I think that it did serve its purpose," Moll says, adding that employees should continue to stand up if the times demand it. "It's a scary thing to do," he says, but "they should feel empowered to speak out when they feel that other avenues become closed or seem closed."

They might have to; since his effort, more revelations of worker dissatisfaction at retail stores have surfaced. In 2014, attorneys, on behalf of retail workers, filed a class-action lawsuit that they said affected twenty thousand employees, alleging they were routinely denied meals during longer shifts and breaks on shorter ones, that they received payments late, and that there were other violations of California labor laws. In 2016, a court ruled in the workers' favor, ordering Apple to pay them $2 million.

At the Beijing Apple Store, Specialists complained of being treated like "criminals," being forced through daily screenings, which they had to wait in line for on their own unpaid time. But generally, worker satisfaction seems high; GlassDoor, the app that workers use to rate workplace satisfaction, shows Apple with high marks. In fact, Apple retail jobs are rated higher than jobs at the company's HQ.

To get a sense of how things might stand for the iPhone salespeople of 2016, I talked to as many as I could. I visited Apple Stores in New York (the flagship glass cube on Fifth Avenue), in San Francisco, in Los Angeles, in Paris (at the Louvre), in Shanghai, and in Cupertino, at Apple's headquarters.

I spoke with dozens of Specialists and Geniuses, none of whom would agree to be quoted by name—Apple's policy of secrecy stretches all the way to the showroom floor.

Generally, people were satisfied with the job; few loved it, few hated it. There was much less of the "cultishness" that critics denounced during the height of i-mania in the mid- to late aughts. Some complained about the lack of flexibility, others hailed the solid benefits. Typical stuff. Perhaps as die-hard enthusiasm—and the once-total secrecy—recedes, along with the shadow of Jobs, that millennial enthusiasm that fueled its once die-hard-loyal workforce will too. But it was an interesting, diverse lot, and I enjoyed chatting with them. I met immigrant jazz musicians and young firefighter trainees and, of course, software developers and part-time repairmen.

There's an Apple Store at Apple HQ, and I popped by after my chat with Apple's PR rep. It's right next to 1 Infinite Loop and one building away from 2 Infinite Loop, which houses the Industrial Design studio, where the very first experiments that would mature into the iPhone were carried out.

Each of the iPhones on the shelf here was designed next door, a couple hundred feet away; the designs were then sent to China, where workers manufactured phones on a massive assembly line and then loaded them onto cargo planes; they were flown to San Francisco and shipped here, to Apple HQ.

As I left the small store, I ran into a small group of Chinese tourists, one of whom asked me to take a photo of them in front of 1 Infinite Loop, which the store abuts.

I snapped the pic and asked the woman who'd handed me the camera why they were here. She flashed a smile and responded immediately.

"We love the iPhone," she said.

CHAPTER 14

Black Market

The afterlife of the one device

You can build anything in Shenzhen from the screws up. It's Silicon Valley's go-to hardware garage. Chips, circuit boards, sensors, casings, cameras, even raw plastics and metals—it's all here.

And if you want to prototype a new product, Shenzhen's Huaqiangbei electronics market is the place to come. I'd heard that you could build a whole iPhone from scratch there, and I wanted to try.

Huaqiangbei is a bustling downtown bazaar: crowded streets, neon lights, sidewalk vendors, and chain smokers. My fixer Wang and I wander into SEG Electronics Plaza, a series of gadget markets surrounding a towering ten-story Best-Buy-on-acid on Huaqiangbei Road. Drones whir, high-end gaming consoles flash, and customers inspect cases of chips. Someone bumbles by on a Hoverboard. A couple shops over, a cluster of kiosks hock knockoff smartphones at deep discount. One saleswoman tries to sell me on an iPhone 6 that's running Google's Android operating system. Another pitches a shiny Huawei phone for about twenty dollars.

I head for a stall manned by a young, shy-looking repairman at work on a gutted iPhone using just a screwdriver and his fingernails, each of which are approximately the length of a guitar pick. I ask him if he knows where to get spare iPhone parts. Without looking up, he nods.

"Can you build me one?"

"Yes," he says. "I think so. But what do you want?" I tell him my model would do; I'm mostly interested in seeing the process.

"It'd be easier to buy the whole thing used," he says. I tell him I'd like to start with the most basic components we can—can we buy the camera sensors, the battery, the boards, and so on individually and put it together, bit by bit? He nods again.

He can make me a 4s for three hundred and fifty renminbi, he says. That's about fifty dollars. And it'd work?

"Of course," he says. I ask if I can record the process, take some photos and video. He calls me crazy, and then, with a hint of trepidation, says sure. He'll throw in a SIM card. Deal, I say.

Without warning, he stands up and takes off. He's cruising—out to the street onto Huaqiangbei Market Road, below an underpass, up across the street, past an upscale-looking McDonald's, down a side street, and into a giant shop space, the insides of which look like an iPhone factory has thrown up all over itself.

In downtown Shenzhen, a couple blocks from the famed electronics market, this smoky four-story building the size of a suburban minimall is an emporium for refurbished, reused, and black-market iPhones. You have to see it to believe it. I've never seen so many iPhones in one place—not at an Apple Store, not raised by the crowd at a rock concert, not at CES. This is just piles and piles of iPhones of every color, model, and stripe.

Some booths are tricked-out repair stalls where young men and women examine iPhones with magnifying lenses and disassemble them with an array of tiny tools. There are entire stalls filled with what must be thousands of tiny little camera lenses. Others advertise custom casings—I'd come back later and buy, for about ten dollars, a "Limited Edition, 1/250, 24 carat gold" iPhone 5 back, complete with the screws I'd need to assemble it. Another table has a huge pile of

silver bitten-Apple logos that a man is separating and meting out. And it's packed full of shoppers, buyers, repair people, all talking and smoking and poring over iPhone paraphernalia.

Our new friend doesn't waste time. He swings by a stall filled with Apple logoed batteries and buys one for fifteen renminbi, about two dollars, and bounds onward. We follow him from stall to stall, watching as he snags a camera module, a black casing, a glass display. We go to a booth with three young women sitting behind it, each staring into their own phones. One wears a white T-shirt with CASH printed on it in block letters. He points to the motherboards below—"That's the whole board," he says. It really is true, you could buy every piece of every going iPhone here.

But I agree to expedite the process and buy a fully stocked iPhone 4s motherboard instead of all the component parts, mostly because he looks a little nervous as I snap photos. Jumble of iPhone innards in hand, we make our exit back to his repair desk at SEG Plaza. He spreads the parts out and sets to work, cradling the device's body in his long fingernails, inserting the battery and the board, and screwing them into place with a custom screwdriver.

Jack, as he tells us to call him, is from a small town near Guiyu, a city built on taking electronics apart. He'd taught himself to repair electronics when he was young, at first as a hobby, just for himself. Then he started to do it for work, and when he moved to Shenzhen, with its massive gadget economy, repairing handsets and tablets made for a good fit.

It's incredible to watch him work—I'd seen the repair pros at iFixit tackle a gadget, and they were impressive. But Jack uses mostly a screwdriver and his bare hands. He is nimble, intuitive, and assured. He assembles the iPhone and all of its components and tests it in about fifteen minutes, and then he hands me my brand-newish, only slightly scuffed iPhone 4s, complete with a SIM card that will let me make calls in China. It feels a little slow next to the 6 I'd been using; otherwise, it works perfectly well.

We celebrate, of course, with a selfie.

New iPhone in hand, we head back to the crowded minimart. There are so many people in the cramped space that standard-decibel conversation fills the room with a roar. When we try to talk to anyone, however, it's suddenly hush-hush—none of them would tell us anything about where they got their iPhones or parts or where they went from here. Some phones are clearly for sale, but many others aren't—vendors wave us off when we ask questions, even about prices. They tell us they aren't for sale—not to us, anyway.

Looks like I'll have to come back and try again. Maybe with an expert.

* * *

"I've never seen anything like this," says Adam Minter when we return to the iWarehouse a couple days later. Minter is an e-waste expert whose book *Junkyard Planet* examines the wide world of tossed, scrapped, and discarded stuff. By luck, we'd both happened to be in Shenzhen at the same time; he was in town to speak at a waste conference.

We wander around the floors, and Wang asks more questions.

Most vendors still refuse to talk, but one thing becomes clear: some of the tables don't sell iPhones to individuals—they're there for wholesale buyers to inspect only.

"Most of these phones are likely headed to Taogao, the Chinese eBay, or eBay, the American eBay." Minter laughs, shaking his head. "Whenever you buy a phone on eBay, you should be wary—it may be coming from here."

Secondhand markets, online or otherwise, are loaded with used iPhones, especially in developing economies like China. It can be big business. But many Americans still think of online markets like eBay and Craigslist as outlets for hand-me-downs. But, as component bazaars like this suggest, it might be part of a larger black—or at least gray—market.

"You know, this makes more sense now," Minter says. A while back, he received a tip about black-market iPhone factories and was able to arrange a visit to one of them. No one would tell him where the parts were coming from. "This, he says, is the missing link." A mass market for every part an operation like that would need to feed an assembly line of recycled iPhones.

Shenzhen has long been known for manufacturing cheap iPhone knockoffs with names like Goophone or Cool999 that mimic the look of the iconic device but could hardly pass as the real thing. But the phones here are identical to any you'd find in an Apple Store, just used.

In 2015, China shut down a counterfeit iPhone factory in Shenzhen, believed to have made some forty-one thousand phones out of secondhand parts. And you might read headlines about counterfeit iPhone rings being busted up in the United States too, from time to time. In 2016, eleven thousand counterfeit iPhones and Samsung phones worth an estimated eight million dollars were seized in an NYPD raid. In 2013, border security agents seized two hundred and fifty thousand dollars' worth of counterfeit iPhones from a Miami shop owner who says he sourced his parts legitimately.

And therein lies the question: What constitutes a counterfeit or black-market iPhone, anyway? The immense popularity of the iPhone, as we've seen, has rippled around the globe, inspiring clones and imitators as well as expansive secondhand markets that buyers turn to in order to get the real thing. Shenzhen's used-iPhone emporium gives us a prime opportunity to consider what makes an iPhone an iPhone—and what happens when we're done with them. There's a reason there's a four-story building in Shenzhen stacked to the ceiling with variations of a single product.

It's one thing if a rogue factory tries to imitate the look and shape of an iPhone and pass it off to unknowing consumers who get home to find out that their phones won't sync to iTunes or that the software is glitchy. But it's really hard to seriously copy an iPhone without, well, iPhone parts. The trademark software and hardware are so tightly integrated, most knowledgeable users would immediately recognize a full-on fake. So, in a sense, any convincing counterfeit iPhone is probably, as far as the user is concerned, an iPhone.

If an iPhone has had its battery replaced, is it not still an iPhone? Or what if the screen isn't made of Gorilla Glass? What if it had extra RAM? Shenzhen phone hackers can jack a phone's memory to twice the amount in standard iPhones. These are all just tweaked, refurbished iPhones, but are apt to be called "counterfeit" by the media. Recall back to the very beginning of our voyage, back in the iFixit lab, how Apple discourages consumers from getting inside its gadgets. Apple uses proprietary screws to prevent tampering, it issues takedown requests on grounds of copyright to blogs that post its repair manuals, and it voids warranties if anyone attempts self- or unlicensed third-party repair. This is probably partly because Apple's repair program nets it an estimated one billion dollars a year, partly because discouraging repairs encourages consumers to buy new, and partly because it prevents the brand from being associated with substandard phones.

Apple does not sell any replacement parts for the iPhone—consumers must pay to have Apple replace things like screens and batteries for them, often at considerable markup. Even aboveboard repairmen can be driven to source parts from used phones, eBay, and places like Shenzhen's black markets. This is why groups like iFixit are pushing Apple and other device makers to ease up on repairers. The issue grew acute enough that in 2016, lawmakers moved to introduce so-called Right to Repair legislation in five states.

Right now, the vast, vast majority of us are not fixing our phones ourselves. When one dies, we slide it into a drawer and buy a new one. Some people throw it out. Others take it into a recycling program.

If your iPhone is still working when you want to upgrade, you have more options. Apple determined that offering trade-in programs would encourage more frequent upgrading, so it launched a Renew program through a wireless distribution contractor, Brightstar. It allows customers to turn in old iPhones for discounts on new ones or for Apple gift cards. eBay makes it easy to resell iPhones, since they retain their value fairly well. And a number of trade-in companies, like Gazelle, have cropped up to solicit old phones for cash.

But what happens once your phone reaches a recycler?

Gazelle and its competitors will first determine if the phones are resalable. If they're in good shape, they might just put them up for sale online. For high-demand items like the iPhone, they might sell in bulk to resellers "around the globe" — for instance, to Chinese companies in Shenzhen that might be able to turn a profit on your two-generations-old iPhone. Since nobody at the Shenzhen market would talk, we can't confirm that. But just as iPhones begin their lives as base elements mined from the earth, often by freelance laborers in barely regulated climates, and are passed upstream through a web of various actors until they end up in Apple's supply chain, they end their lives outside of that network, traded off into increasingly opaque markets. It's just phones and phones and phones. At the black market, it really is.

"I'd guess some of the phones 'fall off the trucks' at factories around here; some of them are from Hong Kong or sourced internationally," Minter says, "and a few of them might come from Guiyu."

Which brings us to e-waste.

* * *

Not long ago, Guiyu was an actual toxic wasteland. Just a few hours west of Shenzhen, it was the Wild West-ish e-waste capital of the world and the site of a serious environmental health crisis. Due largely to its proximity to Hong Kong, which is infamous for its vaguely regulated ports—it's sort of like the Swiss bank account of the shipping industry—Guiyu has become, starting decades ago, a dumping ground for the world's unwanted consumer electronics.

At a stall in a new, half-built complex just off Guiyu's main road, circuit boards, wires, and chips spill out of thin plastic bags, some of which stand four or five feet tall. There are piles of computer guts, monitors, and plastic casings spread out on the concrete. Men and women squat over them, sorting and picking them apart. We walk farther into the industrial complex, where garage doors open to towering walls of still more circuit boards, large and small, the internal kits of desktops and mobile handsets alike.

A man runs out and tells us to stop taking pictures. Another walks by with a wry smile, a stack of circuit boards slung over his shoulders, a lit cigarette between his lips.

To understand why this place exists, we need to go back even farther: In the 1970s and 1980s, as plastic, lead, and toxic-chemical-filled electronics were hitting the consumer market in quantities never before seen, disposing of them became a serious concern. Landfills stuffed with cathode-ray tubes and lead circuit boards (lead solder used to be ubiquitous) posed environmental threats, and citizens of rich countries began to demand environmental controls on e-waste disposal. Those controls, however, led to the rise of the "toxic traders," who bought the e-waste and shipped it to be dumped in China, Eastern Europe, or Africa.

In 1986, one such cargo vessel, the *Khian Sea*, was loaded with fourteen thousand tons of incinerator ash from Philadelphia. The ship sailed to the Bahamas, where it attempted to dump the waste but was turned away. It spent the next sixteen months looking for a place to unload its toxic cargo, trying the Dominican Republic, Panama, Honduras, and elsewhere, and trying unsuccessfully to return it to

Philadelphia before unloading four thousand tons of it on Haiti, telling the government that it was "topsoil fertilizer." When Greenpeace told Haitian officials the truth, they demanded the *Khian Sea* reload the waste, but the ship escaped. The dark tragicomedy of the incident drew international attention, as the ship tried to rename itself—first the *Felicia*, then the *Pelicano*—and continued to court countries to take the remaining waste. Eventually, the ship's captain dumped the remaining ten thousand tons of toxic waste into the open ocean. The ensuing outrage helped spur the formation of the Basel Convention on the Control of Transboundary Movements of Hazardous Wastes and Their Disposal in 1989, which would be signed by 185 nations and ratified by all but, you guessed it, the United States. (And, weirdly, Haiti.) The convention was an effort to prevent what was increasingly called "toxic colonialism" by the victims, and e-waste fell under its purview.

If the ship hadn't been turned away in the Bahamas, this particular winding road of waste subcontractors might never have caught anyone's eye. The city of Philadelphia paid Joseph Paolino and Sons, under a six-million-dollar contract, to dispose of the waste, which that company then handed off to Amalgamated Shipping Company, which was registered in Liberia. At some point in the fiasco, another company, Coastal Carrier, took over operations. The point being, there is a tangled chain of contractors, subcontractors, and foreign companies that make it difficult to track where waste goes after it leaves American homes. For that reason, even today, the Basel Convention remains difficult to enforce.

That brings us back to Guiyu, where similar chains of toxic traders had routed a steady stream of e-waste from around the world to Hong Kong, then to the small Chinese city a couple hundred miles away. This continued well into the 2000s. Recycling and waste-disposal companies in rich countries, it turned out, were offloading their gadgetry garbage onto a site a couple hundred miles from Shenzhen where there's a good chance they were initially assembled.

A Seattle-based nonprofit called the Basel Action Network revealed in 2001 that goods from the United States and Europe had a tendency to end up in the midsize town of Guiyu, where migrant workers were breaking down the gadgets by hand, mixing them in acid baths to remove traces of precious metals, and cooking the circuit boards over coal fires to remove the lead solder. The nearby river ran black with electronic ash, the fields were charred from plastic burning, children were found to have dangerously high levels of lead in their bloodstreams, and miscarriages were rampant.

Today, the driver tells me, they farm rice on the fields they used to burn computers on. It's true that from the main road into town, few of the horror stories seemed to present themselves. As we get closer, there is a large, multicolored billboard trumpeting plans for a new waste-recycling plant. After years of bad press, it appears that the local government is determined to overhaul the town's image.

Instead of letting hundreds of migrants workers burn circuit boards in the open fields, officials created a complex to handle the recycling and metal extraction: an industrial smelter to melt down the wares, and organized stalls that recyclers could rent to more safely break down the electronics. Which is where we'd ended up. The complex was still under construction and only half occupied, though the garages there were packed. This is because, the driver says, many of the former recyclers didn't want to pay rent, so they scattered into more informal operations, behind closed doors, some in town and some on its outskirts. He hints that the government's plan was largely cosmetic, that many of the same activities and their risks persisted, but they'd been swept out of sight.

Which is a pretty good metaphor for the state of e-waste in general. E-waste is an ever-sharper thorn in the technology industry's side. Driven in part by the iPhone-led smartphone boom, which has put complex electronics into more hands than ever before, e-waste continues to be a global blight. For Americans, the lure of dumping it abroad is great—breaking down today's devices is tedious, time-consuming work, and many of the materials aren't valuable.

"There really isn't much in there," David Michaud, the metallurgist who pulverized my iPhone, told me, noting that there's been a lot of talk of recycling the phone for its metals but that it may not be worth the cost of recycling. "You'd need a lot of iPhones to recycle."

In 2016, Apple rolled out Liam, a slick, twenty-nine-armed recycling robot that can rapidly disassemble and sort iPhones into component parts. Apple says Liam is optimized to recycle up to 1.2 million phones a year but nonetheless characterizes the robot as "an experiment" intended to inspire other companies tackling e-waste, and it's unclear what role it will play in the company's long-term operations.

BAN completed another study in 2016; in it, the group teamed with MIT to place GPS sensors in over a hundred electronics submitted to accredited, well-respected e-waste recyclers in the United States, like the Goodwill. Surprise: The majority of them ended up shipped overseas, long after the negative press from exported electronics spurred companies and regulators to try to assert more control over the e-waste-recycling process. Most of the electronics went to Hong Kong. One shipment wound up in Kenya. E-waste-recycling companies that say they're responsibly supervising the landfill-free recycling of American gadgets are still offloading them to China and Africa. Granted, there's a demand for the goods there, where the market for secondhand phones extends even further, and skilled repair workers can revive discarded devices.

"About 41.8 million metric tonnes of e-waste was generated in 2014 and partly handled informally, including illegally," a 2016 UN report, "Waste Crimes," noted. "This could amount to as much as USD $18.8 billion annually. Without sustainable management, monitoring and good governance of e-waste, illegal activities may only increase, undermining attempts to protect health and the environment, as well as to generate legitimate employment."

I didn't see much evidence of good governance at Guiyu—there was no fancy machinery in sight and no protective gear for the workers, who were still breaking it down and sorting it by hand. Instead of squatting in fields and burning circuit boards, they were squatting on concrete and burning them behind closed doors—in a facility we weren't allowed to see—and paying extra for the privilege. We had tea with a local city official, who told us that the plans were not complete and would not be for a year yet.

At least the river wasn't running black, and there were no open flames in sight.

As we drove through town, we saw a building with thousands of tiny microchips scattered out front on the dirty pavement. We stopped, and a crowd of young men in dusty T-shirts looked at us quizzically.

"You want to buy?"

I said sure, I'd take one, picking up the microchip and putting it in my palm. He laughed.

"Keep it."

★ ★ ★

Today, e-waste gathers everywhere, a by-product of the flood of devices, like the iPhone, and the rate they're disposed of. After Guiyu was reined in, reports pegged Ghana's Agbogbloshie dump site as the new "biggest e-waste dump in the world." But Minter says it's the same story everywhere. E-waste flows have grown complex and diffuse, in no small part because the markets for devices have too.

"Honestly, just look at the massive dumps outside of any major city in a developing, less regulated place," Minter tells me in Shenzhen. "Go to Kenya, go to Mombasa, go to Nairobi." Some of the best device-repair technicians Minter has ever seen, he says, can be found there. And the waste dumping is no longer the "toxic colonialism" of yore. Some African and Asian companies are eager to import working secondhand phones. Usually not iPhones, but Android phones and even the cheap Chinese knockoffs will find a second life in African or South Asian markets.

So I decided to try to travel as far downstream as I could. If peeking into a tin mine in Bolivia helped contextualize the origins of the iPhone, perhaps a dump site in a rapidly developing mobile-friendly nation like Kenya could help contextualize its final resting place.

I headed to Dandora, Nairobi's infamous dump, the largest in East Africa. The only way some of the residents of Dandora can get their hands on smartphones is if they dig them out of its churning heaps of decomposing garbage. There are plenty there for the taking too—if you can spot them and root them out. Waste of every kind—from the city and the entire region, from its international airport, and from wealthier countries that have exported their waste—ends up there. Opened in 1975 with World Bank funds, it was declared filled in 2001. But despite city officials repeatedly announcing its imminent closure, some 770,000 tons of industrial, organic, and electronic waste continue to pile up there each year.

The results are predictable—a waste dump that has overflowed for so long that it's become a permanent feature of both the neighborhood it lends real estate to, and the landscape itself.

The smell hits you first, of course; it's the smell of rotting foodstuffs, of spewing methane, of stagnant air and decay.

It's truly massive; hills of garbage roll as far as you can see. Ghoulish, teenager-size storks swoop around scouting for food or stand sentry to the trash.

Three thousand people work in the dump site every day; it's the major job creator for the local economy. They're expert frontline recyclers, and they're looking for everything: basic raw materials like plastic, glass, and paper for recycling; metals like aluminum and copper; and valuable e-waste that can be refurbished and resold. Phones, especially smartphones, are a big draw. If the phones are still intact—many are—pickers take them to the nearby stalls of electronics salespeople; if they're not, they strip out the batteries, motherboards, and copper bits for scrap.

Structures are built directly atop the garbage, which has become a foundation for homes and shops. One building has a skull and crossbones drawn on the door.

"People are born here and die here," Mboma, an actor, clubrunner, and volunteer who was himself born and raised in Dandora, said. I'd met him through friends of a colleague, and he'd offered to show me around. "Some people, all they know is this garbage."

It's a brutal place. The towering garbage hills smolder, emitting noxious gases, and pools of toxic waste collect between them. Those who work here have no protection of any kind and are exposed to the pollution day in and day out. The day I visited, a boy, perhaps thirteen or fourteen, had fallen asleep near the entrance to the site, on the trash-packed road. The driver of the first dump truck to arrive—here they're giant vehicles with conveyor tracks for wheels—didn't see him in time. The truck ran over him, crushing him. Because the Dandora dump site is off the local municipality's radar, police and officials don't visit here or tend to accidents. So the body lay there all day.

My guide told me about the tragedy halfway through our walk around the site; we had passed right by his body at the entrance. There he was, when we returned, covered by a torn piece of cardboard, a pool of viscous blood under his still head.

One of the dump's informal stewards, a young man named T.J., didn't look a day older than twenty-two, but he was apparently a man of some power; the dumps are dangerous but lucrative, so organized cartels control who goes in and out.

"These are good money," he says, bending over and pulling out a basically intact cell phone. He said that they're among the most sought-after items out there. They're often repairable and can be resold in the nearby shops.

I found a Huawei whose touchscreen looked a little melted but could otherwise be usable—T.J. told me screens were hard to repair, since the parts were so scarce and thus my find might not be all that valuable. The full-timers had already found a Nokia body and a frayed BlackBerry. Back in Dandora, such a phone can fetch five hundred shillings (five dollars, but that's a month's rent here), whether it's working perfectly or not.

"Everything is negotiable," says Wahari, a seller who's been hocking wares in Dandora for twenty-five years and is the host of one of the biggest selections of used smartphones in town.

And even here, demand drives a significant market.

"There are two status symbols here," Kinyamu, the entrepreneur, told me. "First, a car. If you can afford a car, you get a car to show you are successful. But second, it is a smartphone."

Indeed, even in Dandora, which many would consider a slum, with its hovels and mud floors and tenuous access to electricity, I see plenty of distracted older youth go by toting smartphones and thumbing screens as they wind through the pedestrian paths, dodging children and sliced-watermelon vendors and crowds gathered outside the packed theater where a soccer game is playing.

It's almost all Android phones. There are a few Apple resellers in Nairobi, but the iPhone is still a luxury here, where it's well known, but rarely seen.

"An iPhone is the ultimate status symbol for a businessman to bring into a boardroom—actually, now it's an iPad."

Wahari, the Dandora recycled-goods salesman, says once in a while they'll find an iPhone out in the dump site.

"Oh, it's very rare," he says with a laugh, and he shakes his head. "Very rare. But it happens, and that is a good day. It is the gold mine."

★ ★ ★

There are few places around the globe that remain untouched by the influence of the iPhone. Even where it's an aspirational device, it has nonetheless driven mass adoption of smartphones built in its likeness and kindled, yes, a nearly universal desire, as Jon Agar put it.

Now, a last step remains before we can successfully reassemble that gold mine and understand that one device.

Once all of the parts and pieces we've explored in this book had been laid out, so to speak, in various places around the world, those materials and technologies, they had to be pinpointed, collected, improved, and artfully innovated by Apple.

This is the story of how that finally happened.

iV: The One Device

Purple reign

The Purple Dorm, aka Fight Club, aka 2 Infinite Loop, second floor, was packed. The aging office space—Apple's HQ was built in the early nineties, and purple and teal accents dotted the halls—had become a hub of activity. The conference rooms in the wing were cheekily named—Between, Rock, and Hard Place, for example. Another was called Diplomacy, which was where Christie's crew banged out the new UI. Fishbowl was the main conference room, where Steve Jobs was a weekly presence.

By 2006, the basic contours of the iPhone project had been defined. Members of the Mac OS team and the NeXT mafia would engineer the software; the Human Interface team would work closely with them to improve, integrate, and dream up new designs; and the iPod team would wrangle the hardware. A team was working day in and day out to identify and strip lines of code from Mac OS to make it fit on a portable device. The famed ID group had set about perfecting the form factor. And Bas Ording, Imran Chaudhri, and Greg Christie's old office space had become the gravitational center of the project.

"It was like, 'Oh, this is going to be real?'" Ording says. "'Now, how do you actually go through all your photos? How does mail really work? How does the keyboard work exactly?' So you have a ton of stuff to figure out." HI had established the basics of what the phone was going to look and feel like—a powerful enough vision to supplant the less risky, but less cool, iPod phone.

"From the beginning, it was all about trust," Chaudhri says. "People thought computers were too complicated, even Macs. So I was really designing interfaces that my father could use. We were trying to create systems that people could use intuitively, that they could trust." So far, they had succeeded.

"It was totally logical," says Henri Lamiraux, the man who led the software-engineering effort under Forstall. "The HI team did a good job of creating a mock-up of the UI. We had a good idea of what it was going to look like and how it was going to interact."

Lamiraux is one of the most universally respected engineers on the iPhone team—his calm, even-keeled style often helped steady others when crises hit. He's the opposite of a stereotypical Apple boss; he speaks in a lilting French accent, boasts a stubble-white beard, and seems more like an abstract-art sculptor than an engineer. Then again, abstract sculpting was a big part of the job. "This thing evolved quite a bit compared to what we thought, but the spirit was already there," he says.

"What you see as the Springboard with all the icons, that was there from day one," Lamiraux says. "And the dock, that was there from day zero." Lamiraux and his team figured out how to code the ideas that had been cast on a lumbering prototype, translated to a tablet, and, now, downsized again for smaller devices.

The P2 crew called those tethered units "wallabies," and they were the go-to tool for experimentation in the Purple Dorm.

There's a reason that all those software engineers had migrated to the interface designers' home base—the iPhone was built on intense collaboration between the two camps. Designers could pop over to an engineer to see if a new idea was workable. The engineer could tell them which elements needed to be adjusted. It was unusual, even for Apple, for teams to be so tightly integrated.

"One of the important things to note about the iPhone team was there was a spirit of 'We're all in this together,'" Richard Williamson says. "There was a ton of collaboration across the whole stack, all the way from Bas Ording doing innovative UI mock-ups down to the OS team with John Wright doing modifications to the kernel. And we could do this because we were all actually in this lockdown area. It was maybe just forty people at the max, but we had this hub right above Jony Ive's design studio. In Infinite Loop Two, you had to have a second access key to get in there. We pretty much lived there for a couple of years."

A feature called Jetsam, he says, is a good example of what could happen as a result. They needed to come up with new ways to parcel out the device's precious memory if they hoped to make the iPhone as fluid as Bas and Imran's demos. Williamson proposed the Jetsam concept, which would terminate unused applications that were draining too much memory. An OS engineer named John Wright took it on.

"Because everyone was right there, I could take a crazy idea like Jetsam, I could go talk to John and say, 'Look, is this crazy or can we actually do it?' And he'd say, 'Yeah, it's crazy, but we can probably do it.' And then Bas would come by and say, 'I want to do this crazy animation, could we do that?' And we'd say, 'No.' And then John Harper would say, 'Well, we could probably do that.' That's one of the things that was special. Everybody there was brilliant."

"That project broke all the rules of product management," a member of the original iPhone group recalls. "It was the all-star team—it was clear they were picking the top people out of the org. We were just going full force. None of us had built a phone before; we were figuring it out as we went along. It was the one time it felt like design and engineering were working together to solve these problems. We'd sit together and figure it out. It's the most influence over a product I've ever had or ever will have."

That tight-knit team wasn't just packed in—they were sealed off. This was Apple's version of Fight Club, after all. "This is one of the things that Steve has done brilliantly," Williamson says, "this idea of building what really is a start-up inside a larger company and insulating it from everything else that's going on in the company. And giving them essentially infinite resources to do what they need to do."

Here is what the iPhone software start-up's org chart would look like: Steve Jobs is CEO of iPhone Inc. Reporting directly to him is Scott Forstall, the head of iPhone software. Under him there's Henri Lamiraux, who oversees Richard Williamson and Nitin Ganatra, each of whom manage small teams of their own—Williamson did Safari and web apps, and Ganatra did mail, the phone, and so on. Also under Forstall is Greg Christie, the head of the Human Interface group, which includes Bas Ording, Imran Chaudhri, Stephen LeMay, Marcel van Os, Freddy Anzures, and Mike Matas. Also reporting to Forstall is Kim Vorath, the product manager who will come to oversee the quality assurance department and who will be one of the only women on the iPhone's software team.

All told, between design and software engineering, there were twenty to twenty-five people working on the iPhone in its early stages—a paltry number, given the known stakes and the ultimate impact of the device. Forstall was a constant presence, and Jobs was given regular demos of the progress. "It was the most complex fun I've ever had," Jobs said. "It was like being the one evolving the variations on 'Sgt. Pepper.'"

"The main process was very interactive," Lamiraux says. "There was a weekly meeting with Steve, so we had a list of features, a list of things we had to get approved. So half of it was the HI team showing some mock-up of what they thought our feature should look like, so Steve was approving, so Steve would say, 'Ah, I like A,' so my goal was to take A and make it happen. And the next meeting, we'd say, 'A is implemented, what do you think?' And he'd say, 'Oh, it sucks, let's try B.'"

You've Got Mail

Nitin Ganatra was among the first to receive one of Scott Forstall's famous recruitment visits.

Ganatra was born in 1969 in Vancouver. Like many of his iPhone peers, he proved adept at computers early on, and he learned to code while still in elementary school; he wrote a program to help him learn Spanish. He'd been at Apple since the early 1990s, had weathered the dark years, and was leading the team behind Apple's mail client.

"Scott came into my office and said, 'There's this effort taking place,'" Nitin Ganatra says. "We're actually going to work on a phone. We have some designs that are done, and we need to start prototyping and figuring out how we're actually going to ship this thing." In other words, it was go time.

"Email was a big function of these phones," Ganatra says. "We saw that from the BlackBerry. So we knew that we had to nail email. I think that was something that was big on Scott and Steve's minds: We can't come out with a smartphone and try to take on the king of email and not have a great client ourselves."

After accepting the project, he was led into the dank, windowless room that housed the touchscreen prototype. "The very first reaction was amazement," he says. "It was probably similar to what a lot of people felt when they first saw the phone. Just, 'Yes, this is what I want. I want this in my pocket right now; how can I get this in my pocket right now?'" The glee would be short-lived, however. "That very quickly turned into 'Holy shit, how are we going to make this run all day?'" Ganatra says. He pauses. "I guess maybe as an engineer that tends to happen. You have more questions than answers almost all the time."

The iPhonic Ingredients

One former iPhone engineer, Evan Doll, reckons there are two unique components that helped the iPhone excel.

There were "these two pieces of tech, which were each basically created by one person." One was Wayne Westerman's FingerWorks. "And that single-handedly was the genesis of multitouch." And then there was John Harper, an engineer whom Doll describes as "a pretty hard-core introvert," who created Core Animation. "Which was the foundation for doing these really fluid, animated user interfaces that Google, still, however many years later, with Android has not really caught up with."

It's a compelling case — multitouch, of course, was the innovation that Jobs seized on in the keynote demonstration. But Core Animation is the framework that allowed developers to bring multitouch to life — touch an icon, and it immediately dances below your finger.

Core Animation works by "handing most of the actual drawing work off to the onboard graphics hardware to accelerate the rendering," according to Apple. It's an ultra-efficient way to ensure that apps can run attractive animations, and it would let developers tackle lively apps with ease. "John Harper is the genius behind that," Williamson says. "It's one of the things that made the first iPhone, which had very little computing power, perform well."

On the back of Core Animation, the Purple team was in the process of enshrining the interactions that would rise to cultural dominance. Some had already been imagined, inside Apple and out, and needed to be executed. Some needed to be dreamed up altogether. For instance, the P2ers needed a way for users to signal they wanted to activate the device, without relying on hard switches, which Steve Jobs despised. Once it was turned on, the screen would have to go dark, so the phone could be ready to receive calls without the battery draining away. Tapping the Home button would wake up the phone, but that could accidentally happen in a user's pocket — again, risking serious battery drain. So the designers needed to come up with a software hack that would be simple for users to do — with one hand, ideally — yet complex enough to prevent accidental activations.

Imran Chaudhri had an idea to rotate your fingers over the screen like turning a knob, but it felt a bit too complicated. Chewing over the problem, the UI designer Freddy Anzures, who'd been working on the unlock concept with Imran, took a flight out to New York from San Francisco. The team had been thinking about how to open the phone, and, well, there he was: He stepped into the airplane bathroom stall and slid to lock. Then, of course, he slid to unlock. That, he thought, would be a great design hack. It was a smart way to activate a touchscreen whose sensors always needed to be on.

Later, Chaudhri had an idea to test the concept out. He had a baby daughter at home, and he placed a prototype in front of her. When even she was able to slide to unlock, he knew it'd be universal. Ideas like that were spilling forward from the design team and the software engineers. The process was open and multifaceted; just look at the patents.

"Those patents, you see a lot of names on them...it was a very small team, and we were all working together. So we were having a lot of discussion," Lamiraux says. "It's not like someone went to his office one morning and says, 'Okay, I'm going to have an idea today, hmmm, let's do visual voicemail.'"

Some features were simply mandatory. As much freedom as the carrier was willing to give Apple, the iPhone would have to abide by certain requirements. "Cingular had a list of features we had to have," Lamiraux says. "So we had to have voicemail, because that's what a phone was supposed to do. So we said, 'Okay, voicemail, but you know, we want to do something better — so how can we make it better? Well, what if voicemail was like email?'"

Ideas bled through the group, and the members seized on them, implemented them, tested them, discarded them, embraced them. They emerged in brainstorming sessions, in late-night coding sprees. In fact, Lamiraux says that he can identify only one design idea as his own among them: "I don't know if you notice, on the iPhone when you open a window, you see the scroll bar flashing, the scroll bar on the side flashes to show that you can scroll. And I will always remember the day when I came up with that, because we are in a meeting, and it was, 'Okay, how can we show people that there is something to scroll?' I said, 'Why don't we flash it?' They said, 'Okay!' That was it."

As the ideas came forward, one thing was fast becoming clear: as promised, this was going to be a colossal amount of work.

"Take the email client, for example. Oh, yeah, here's the list of your messages and you tap and see how it opens, and it's cool," Ording says. "But, oh, wait, how do you reply and how do you forward and how do you get multiple mailboxes, and all of a sudden there's a ton of stuff you have to resolve to make it really work as a full mail client. And the same for voicemail. It expands really fast. And you discover just how much you have to resolve to make it work properly."

Yes, the iPhone Was Inspired by *Minority Report*

The touch-based phone, which was originally supposed to be nothing but screen, was going to need at least one button. We all know it well today—the Home button. But Steve Jobs wanted it to have two; he felt they'd need a back button for navigation. Chaudhri argued that it was all about generating trust and predictability. One button that does the same thing every time you press it: it shows you your stuff.

The story of the Home button is actually linked to both a feature on the Mac and everyone's favorite science-fictional user interface—the gesture controls in *Minority Report*. The Tom Cruise sci-fi film, based on a Philip K. Dick story, was released in 2002, right when the ENRI talks were beginning. Ever since, the film has become shorthand for futuristic user interface—the characters wave their hands around in the air to manipulate virtual objects and swipe elements away. It's also an ancestor of some of the core user-interface elements of the iPhone.

"*Minority Report* was very cool stuff, very inspiring," Ording says. "You know the Exposé feature?" Exposé is a feature Ording wrote for the Mac that's still a core part of the UI today—it allows you to zoom out on all your open windows so you can take stock of everything you have open at once. "I was staring at my screen with a whole pile of windows, and I'm like, 'I wish I could somehow, just like they do in the movie, go through, in between those windows and somehow get through all your stuff.' That became the Exposé thing, but it was inspired by *Minority Report*." And Exposé, in turn, would inspire a core functionality of the iPhone.

"I remember for the Home button, Imran, he worked on some early ideas for that, like, that there was a button, and he originally called it 'Exposé for the iPhone.' So that you have one button to see all of your apps. And then you tap one and it zooms in on that app, just like you choose a window in Exposé. And then later, that became Menu, or Home."

"Again, that came down to a trust issue," Chaudhri says, "that people could trust the device to do what they wanted it to do. Part of the problem with other phones was the features were buried in menus, they were too complex." A back button could complicate matters too, he told Jobs.

"I won that argument," Chaudhri says.

* * *

Creating features was one major task. Refining the experience of using them was another.

"There were these known truths that we discussed," Ganatra says, "and [that] we knew couldn't be violated."

1. Home button always takes you back home.
2. Everything has to respond instantly to a user's touch.
3. Everything has to run at least sixty frames per second. All operations.

On top of all that, the experience itself had to be fine-tuned in every arena. For instance: "There was an awful lot of work that went into the acceleration and deceleration curves for scrolling," in nailing the physics of swiping through lists. Jobs and Forstall drove the team hard on that point. "The goal had always been to make the iPhone feel like you were touching something real. Steve and Scott really wanted the interaction to be, you push something and it moves. No delay," Lamiraux says. "You felt like you were touching a piece of paper, and it was scrolling under your fingers."

That natural physicality extended to the design of the apps. "There was a lot of work that went into mimicking physical and familiar things that people were already used to interacting with," he says. And that's where the iPhone's infamous skeuomorphism—the designing of digital objects to resemble versions of real ones—came in.

"Early on, skeuomorphism was one of the things that made it so that people actually understood how to use an iPhone when they picked it up—there were already physical things in their life that they could model their interactions after, and that gave them clues as to how to use the device," Ganatra says. "It really did start as, 'Let's try to model these things after things people already know how to do.' And that was already happening on Mac OS X. In fact the previous apps I was managing, Mail, it had a postage stamp for the icon, and there was an address book, and it was a book, and so on. We knew we didn't want to have anything like a user manual. If you ship one of those, you've already kind of failed."

Carrier Me Home

The iPhone would be billed as three devices in one—a phone, a touch music player, and an internet communicator. So it was important that the internet portion was well accounted for. "I had this belief that the web was fundamental to how we were going to be interacting with mobile devices, so my perspective was that the web was important too, in addition to the phone and the music player—in fact, probably more important than the other two," says Williamson, who was in charge of porting the Safari web browser to iOS.

At the time, the standard protocol for the mobile web was WAP, or wireless application protocol. In order to limit the use of wireless data, WAP allowed users to access stripped-down versions of websites, often text-only or with only low-resolution images.

"We called it the baby web—you got these dumbed-down web

pages," Williamson says. "We thought that it was maybe possible to take full-on web content to display in one place." At the time, few if any smartphones or mobile devices allowed users to browse the proper web. Carriers saw data plans as their future but pushed restrictive pay-as-you-go data plans that were prohibitively expensive and ignored by the public.

As the software and hardware teams scrambled to make a working phone, the negotiations with the carriers—first with Verizon, and then with Cingular (soon to become AT&T)—had been going on in the background. They concluded in 2006, with AT&T winning out, albeit with some important concessions. "I was involved to the extent that what I wanted to argue for was a transparent data pipe," Williamson says. "Which was something the carriers had never given anybody. Up until that point, WAP was predominant."

Carriers were favoring the conditions of the network over the device, trying to ensure speed over quality. "The carriers used to actively filter content, so they would do things like, if you were trying to show an image, they would transcode the image to lower the resolution so it'd be smaller so it'd be a little faster to the device," Williamson says. "So we went through a lot of negotiations with AT&T to get them to agree to a clear pipe. And now it's become de facto in all contracts." They also had to negotiate with AT&T to get persistent connections. "Without a persistent connection, you can't do things like notifications, and things like iMessage become a lot more difficult. They said, 'No, we can't do persistent connections! We have millions of devices! No, no, we can't do that!'"

They did that. "That went into the contract too. We wanted to bring them out of the mentality where, you know, they were in, which was, 'We want you to pay for ringtones, pay for text messages,' and into this reality of 'It's just a computer that needs an IP connection, and we want that and everything you can get with an IP connection.' So that, that's huge, in terms of enabling the device to actually be capable as a modern smartphone."

It's hard to overstate how crucial a development this was. If you can remember using pre-iPhone cells for anything besides making calls, you also remember cascading phone bills filled with text-message charges and ringtone and game-download fees. AT&T was forward-thinking, but they were also very concerned about their business model, Williamson says. "In fact we eviscerated them. Nobody wants to pay for text messages or ringtones anymore. But they also did really well as the exclusive carrier for the iPhone. So it was win-win."

Meanwhile, Jobs's fears of the carriers sending them big books filled with specifications had come to pass. Williamson remembers interminable meetings with AT&T and their technical folks about issues. "They would come with these specification books, and they were like, 'We have to do it this way, we have to support that.' And we were like, 'No. No. No.' And we triumphed in the end; it was just hard to get there."

For an idea of how central the negotiations were to the project, consider that Apple hired a project manager specifically to oversee carrier relations. "At one point, his team was the same size as the software team," Williamson says.

Hardwired for Touching

And then, of course, there was the hardware. Tony Fadell started hiring hands from around the company and, because the extreme-secrecy clause applied mostly to the user-interface and the industrial-design teams, hiring new engineers and third-party suppliers.

"We had to get all kinds of experts involved," he says. "Third-party suppliers to help. We had to basically make a touchscreen company." Apple hired dozens of people to execute the multitouch hardware alone. "The team itself was forty, fifty people just to do touch," Fadell says. The touch sensors they needed to manufacture were not widely available yet. TPK, the small Taiwanese firm they found to mass-manufacture them, would boom into a multibillion-dollar company, largely on the strength of that one contract. And that was just touch — they were going to need Wi-Fi modules, multiple sensors, a tailor-made CPU, a suitable screen, and more. The list was exhausting.

"Any one of those was very difficult," Fadell says. "All of it together was a moonshot. It was like the Apollo project."

Thanks to the ENRI team and the FingerWorks crew that Apple acquired in 2005, it had the know-how. "We had the basic science. It was about having the right technologists for chips. The right technology about manufacturing.... The question was, could we scale it and make it work with all the different environments," Fadell says. For instance: "We had a real problem with sweat; that would make it fail. And we switched from plastic to glass at the very last minute, which was a curveball."

Because they were trying to run Mac-caliber software on a tiny device, the hardware constraints were considerable. "Everything had to be really well fucking optimized," Grignon says. "So we built our own chip."

And Grignon, who was the senior engineer in charge of radios, had something of an outsize job. "Introducing radio into a handheld device was something we had never really done," he says. They had experimented with the iPod phone, but they'd never attempted to scale into something that would be ready for the mass market. "We had to, because of the enclosure materials, we had to engineer our own antennas, which are its own set of art and magic." It would take an enormous effort to build and test those antennas to make sure that they'd work.

Put it all together, Grignon says, and you've got a recipe for madness: "At the fundamental hardware layer, everything is new."

A brand-new CPU running a brand-new operating system running brand-new apps interfacing with brand-new hardware. "Imagine you're a developer or a tester, and you'd have a crash," Grignon says. " 'Oh, shit, the app crashed. Why'd the app crash?' Well, it could be at any layer in that entire stack, all the way down to the silicon. I mean, think about that. Imagine an entire piece of silicon that can shit the bed, because it's new. We would have actual CPU bugs, or compiler bugs, because we were building the operating system for a different instruction set. Or we could have an actual legit bug in the coding, a logic error in the app. It was just a fucking nightmare."

Key Us In

For a moment, there was a chance that the iPhone would kill QWERTY.

"The radical idea was that we had no physical keyboard," Williamson says. "In hindsight, that was obviously the right thing to do. But then, we were all very concerned." BlackBerry was finding success with its hard-button keyboard. And fear of another Newton-style input misfire was thick in the air. "We all had this fear of, you know, the Newton disaster," Ganatra says.

In the 1990s, the Newton's glitchy handwriting software had been so widely derided that an episode of *The Simpsons* took aim at the device. One bully tells the other: "Take a memo on your Newton: 'Beat up Martin.'" The device can't read the input and spits out, "Eat up Martha." So "Eat up Martha" became a cautionary mantra among the engineers, repeated often in the Purple Dorm.

To ensure that people using the device would be able to interact with the objects on the screen accurately, the engineers determined a "minimum-region hit size," or how small something could be while still reliably responding to clumsy fingers. "Anything that you touched on the screen had to be as big as that size, otherwise it was too hard to use, and you'd make a lot of mistakes," Williamson says. But given the size of the phone's screen they were working with, a QWERTY-style virtual keyboard was out of the question—the key buttons would be too small. "So we had this big conundrum. In fact, the early prototypes we had were terrible."

It kept triggering wrong key presses. "Initially it didn't work that well," Ording says. "So it was a good time to rethink how you enter text. Because QWERTY is based on old typewriters, and it's a little strange. On the other hand, people know how to use it. So that's why there were other experiments, lots of exploring there."

QWERTY, the keyboard layout named for the order of letters from the top left over, was literally built for inefficiency. It was designed to keep nineteenth-century typists from hitting the keys too fast and jamming up early typing machines. It's persisted for over a hundred years because of its familiarity—people who knew how to type on typewriters could easily transition to computer keyboards, and so they did. The layout has lingered on despite the suggestion of more efficient configurations, like DVORAK, that have all but proven to increase typing speed. The prospect of a new, key-free touch surface opened the possibility of reimagining how we input text and offered a chance to break away from a centuries-old layout.

A good keyboard, of course, was make-or-break for a device that relied on text input for the vast majority of its functions. So the designers and engineers had to get creative.

"We kind of took a hiatus from general development," Williamson says. "And we encouraged anyone who wanted to to write a keyboard. It was kind of a fun time; we'd all been super stressed out, and getting the freedom to do anything you wanted, knowing it wasn't going to have to ship, was a good diversion. There must have been a couple weeks we did that, which doesn't sound like a lot of time, but for this group it was."

Some engineers proposed chord keyboards, which would divide the screen into a 3x3 grid, and users could select a letter by touching two of those regions. "We had bubble keyboards where you could click and slide," Williamson says. "You'd click on the screen, and you'd get a pop-up with four letters, and then you'd slide onto the letter you wanted."

New algorithms were tested and altogether new layouts were examined. Radical rethinkings of text input were floated.

"A lot of people were thinking, 'We can do whatever,'" Ording recalls. "Things where you can swipe on the keys, or double tap, or there were a whole bunch of variations. There were different orders based on frequency of key use and letters." New layouts that might take a while to learn but would ultimately prove more efficient. "We tried all kinds of stuff to come up with all kinds of variants to make keys appear bigger or have a multitap that you could use to cycle through the letters."

"The chord keyboards were probably the most crazy," Williamson says. "One of them was like a piano keyboard, and you could kind of play letters on the keyboard." Another keyboard closely resembled Graffiti, the much-maligned input technology on the ill-fated Newton. "We had a Graffiti-esque keyboard that got shot down very early."

The team put a website together to compile the keyboard designs. An engineer named Ken Kocienda "won" the contest and ended up leading the keyboard project.

By Williamson's estimation, they developed around half a dozen alternative entry methods. They went so far as to design ways to introduce the brand-new keyboard layouts—which, naturally, would be totally alien to users—through simple learning games. Williamson says, "We went down this path of thinking we'll ship the phone with this keyboard game. That will teach you how to use the keyboard. Some of the games were, like, you had to type the letters in a certain time frame, it counted down, or you could blast letters in a word—fun little games." Jobs, however, was not amused.

"We showed Steve all of these things and he shot them all down. Steve wanted something that people could understand right away," Williamson says. They stuck with the suboptimal key configuration for the same reason it had migrated to computers half a decade ago—familiarity. "When people pick up this phone in the store, it has to be something that's instantly recognizable, that they can use immediately. And that's why we stuck with the QWERTY keyboard, and we added a whole bunch of smarts in there."

Those smarts would be crucial. "People thought that the keyboard we delivered wasn't sophisticated, but in reality it was super-sophisticated," Williamson says. "Because the touch region of each key was smaller than the minimum hit size. We had to write a bunch of predictive algorithms technology to think about the words you could possibly be typing, artificially increase the hit area of the next few keys that would correspond to those words."

When you hit a letter, the predictive software guesses what you're going to hit next, and it enlarges your minimum-region hit size. So if you tap *H*, the hit size around *I* and *E* widens, rendering the keyboard more forgiving. With the help of those algorithms, the keyboard improved.

There's another way to gauge the importance of the keyboard: it's the only part of the entire project that was user-tested outside of the core team.

"We were so worried about the accuracy of the keyboard that we had everybody that was disclosed on the phone but not working on the software do usability testing," Williamson says. The once-abandoned user-testing lab where the sapling of the iPhone was planted would finally serve its intended function.

The First Rule of Fight Club

The locked-down Purple Dorm was bustling. "There was never enough time, never enough people," Lamiraux says. "People worked very, very hard."

They were adding people to the team, but slowly, largely because of the secrecy demanded by Jobs and Forstall. The UI was their crown jewel. No one was allowed to see it unless they were on the P2 team or had received explicit approval from Jobs. Initially, that was only a small number of Purplers. "Less than fifteen or twenty people tops could see the UI, including the UI designers that were drawing the pixels," Grignon says. As a member of the hardware crew, he was initially barred from seeing it too.

If Jobs wasn't around, P2 couldn't add any engineers, even from inside Apple, even if they wanted to. Management referred to those who'd been approved as "UI-disclosed."

Inside the Purple Dorm, the engineers were too busy to think much about the security measures.

Outside, however, the obviously cordoned-off building broadcast a certain exclusionary vibe to the rest of the company. "It was literally locked down with a metal door, which is bizarre and unsettling," one iPhone team member says. "Steve loved this stuff," Grignon says. "He loved to set up division. But it was a big 'fuck you' to the people who couldn't get in. Everyone knows who the rock stars are in a company, and when you start to see them all slowly get plucked out of your area and put in a big room behind glass doors that you don't have access to, it feels bad."

If engineers outside the Purple team were called in to debug technical issues, black cloth would be draped over any screen that might be displaying the user interface. "When you have engineers who are separated with a cloth between them debugging a problem, that's dumb," Grignon says. Taking that to its logical conclusion, the AT&T people could never see the phone either, and they didn't, he says. "Never. They saw it onstage when we announced it along with everybody else."

Then there was Steve Jobs.

"The fear of Steve was great," Evan Doll tells me. "He was feared more than anything else by the rank and file, and even the middle-management layer at Apple. It was like a cult of personality. He would come walking down the hallway, and I would shut the fuck up," Doll says. "People were more concerned about the downside of a Steve interaction than the potential benefit. I don't want to make it seem like it was this gulag environment, but there was definitely a strong undercurrent of fear, paranoia, that definitely was a part of any interaction that a team might have with Steve, for sure."

Jobs did look for ideas in parts of the company outside the Purple Dorm—he just didn't tell anyone what they'd be working on. Abigail Brody was the creative director heading up the Pro Apps group. She was handed down a request to work on a mysterious project, something called "P2." "They told me, 'You have to work on a multitouch project,'" Brody says. "And they gave me a multitouch prototype that was a little bit smaller than an iPad, but it was bigger than a conventional phone. It was a very crudely put together thing. If I remember correctly it was taped, so I could experience the gestures." They wanted her to design a user interface and a health management app, among other things. But her team was told little else. "We had no clue whatsoever," she says. "We were just told, there's a list view and a main menu, we need a gallery and this and this—it was pretty vague. The only thing they did not mention was a phone."

Meanwhile, third-party suppliers tapped to work on various iPhone components would be given false schematics so they'd think the project was just another iteration of the iPod. Member of the iPhone team would pose as representatives from other companies when meeting with vendors to avoid starting rumors. And everyone had to sign strict nondisclosure agreements that stipulated they could be fired if they leaked information about the phone.

"That whole experience was, like, you're a ninja, you don't exist," one iPhone designer says. "It was weird, samurai-type shit."

Sometimes, new recruits had to sign a preliminary NDA first, agreeing that they would never discuss the existence of the next NDA they were about to sign, in case they didn't want to sign that one. "It's always weird if you have to try to be all secretive about things," Ording says. In the beginning, he was told to keep the project secret even from the rest of his own HI team.

The closed doors were demoralizing some employees and agitating others. Especially those on the iPod team who were, essentially, tasked with building the hardware for a device whose software they weren't allowed to see. They had to make a fake operating system so people could actually test it out.

"That's when we invented skankphone. That was Apple's kind of exclusivity that was a mix of paranoia and politics at its worst," Grignon says. "Skankphone was just like a clown vomited all over the screen. It was the worst-looking dialer, it texted, it had all the functionality, but that's what the quality assurance people could use. It still was built on iOS at its core, but none of the actual UI widgetry was there. It allowed the AT&T people to test it, our own QA people, but, you know, eighty percent of the people on the program couldn't see the actual UI that it was going to ship with."

By insisting that the UI remain secret, Forstall made life difficult for Tony Fadell, who had recently been promoted to senior vice president and who was the only one on his team allowed to see the software. "Forstall skillfully played that off of Steve; he fed the paranoia to keep it super-secret, to keep Tony out of it," Grignon says.

That secrecy was bleeding into the general culture at Apple, creating a wedge between friends and impeding actual progress on the phone.

Nitin Ganatra and Andy Grignon were and remain good buddies. When I was doing a batch of interviews around Silicon Valley, I met Grignon in Half Moon Bay, where we chatted at a seaside pub. The next day, I met Ganatra for lunch at a Mexican joint in Palo Alto. I told him that I'd seen his old friend Andy Grignon the day before, and he had given me a message to deliver. Ganatra cut me off.

"Did he tell me to go fuck myself?" Ganatra says, grinning. He had indeed. "Yeah? Well, next time you see him, make sure you tell him the same." See? Friends. But here's what Grignon told me about working together under the iPhone era:

"People may be best buddies off the field, but on the field, it's anything goes. And I may have to do some dirty things, and that's horrible. So there were moments where, you know, Nitin and I are close friends to this day, outside all of the bullshit, I count him as a really close friend. I love hanging with him—but at the time I wouldn't have thought twice about throwing him under the bus, or if there was an opportunity where I thought it could help us to break his balls a little bit or make his life a little bit harder, just to apply some pressure, I would have done it. And he did the same to me, and I know that. But then we go and drink and smoke and do whatever, and it's all good."

Before Grignon left to join the iPod team, he had worked closely with Ganatra, Scott Herz, and other members of what was now the Purple team. They used to eat lunch together—"We'd gripe about whatever the fuck we were working on, Mail or iChat, and we'd break balls or whatever, that was cool"—but over the course of the iPhone development, the mood changed.

"Lunch became one of those Mexican standoffs. We would still go through the ritual of lunch. And they would talk in code names and go, 'What do you think about XYZ,' and it would be some code name. And I would be like, 'What's that?' And they'd be like, 'We can't talk about it,'" Grignon says. "It got really weird there for a while. Very passive-aggressive. It depends on who can withhold what information...And sometimes we'd just sit there and eat real quiet. Idle chitchat, but so obviously we couldn't wait to get back out of the environment we were in. And these were also my friends. I'm not supposed to be honest with my friends? It was really fucking weird."

To this day, Tony Fadell sounds exasperated when the conversation turns to iPhone politics. "The politics were really hard," he says. "And they got even worse over time. They became emboldened by Steve, because he didn't want the UI—I could see it—but he wouldn't let anyone else on the hardware team see it, so there was this quasi-diagnostics operating system interface. So it was super-secretive, and it emboldened the other team...You had to ask permission for everything, and it really built a huge rift between the two teams."

The team that was building the iPhone's hardware and the team that was designing the software were distinctly at odds. "The teams didn't want to work together. Or they just wanted to blame each other," Fadell says. "And it's like, no, that's not how it works." It's a pretty remarkable way to build a product, especially one in which the hardware and software are so tightly and powerfully integrated. Eventually, the secrecy made it too difficult to make any meaningful progress."

"It got to the point of absurdity," Grignon says, "where I was like, we couldn't make progress, and we were moving slow because we couldn't work with the actual UI, so Tony had to go directly to Steve and be like, 'Look, I need Andy to see the UI.' And Forstall argued, but then gave in. Tony was able to successfully negotiate that path and say, 'We can't build a fucking product if at least some of our close people can't see it.' It was absurd."

The move pushed Jobs to allow five or so more people to the UI-disclosed list, and, amazingly, Scott Forstall himself used it as an excuse to grant access to a number of people on his own team who didn't have access. The secrecy was out of control, even for Apple, Grignon says, and ultimately detrimental to the project.

"Oftentimes, it's just hidden under 'Oh, Apple being secretive again; oh, those guys!' but it was stupid," Grignon says. "Even in that Apple-rarefied space of paranoia, it's still stupid, and that's where politics come in. Can you have good products without politics? I would say you can. I think some politics are good. But it does burden the development process unduly. You do more working together."

Industrial Design

The Industrial Design group had been involved in the inception of the phone nearly every step of the way: Duncan Kerr was an influential participant in the ENRI sessions, ID was responsible for the form-factor designs that helped skyrocket the iPod to popularity, and executives like Mike Bell had used a batch of ID's prototypes as an argument that an Apple phone could succeed in the first place.

It's fitting, then, that the very first known design sketch Jony Ive made of a touchscreen very closely resembles the iPhone screen that actually ended up shipping.

"Some of our early discussions about the iPhone," Ive said, centered on the idea of "this infinity pool, this pond, where the display would sort of magically appear." From the earliest talks, the emphasis was on elevating the screen; as he put it, everything should defer to the display.

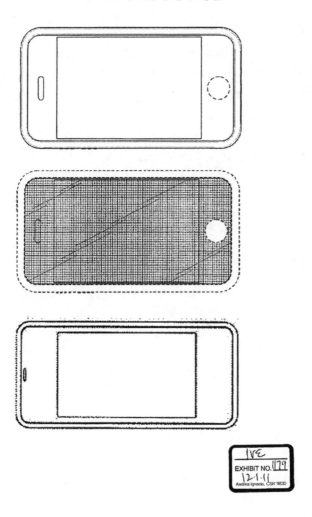

EXHIBIT NO. 179
12·1·11
Andrea Ignacio, CSR 9830

Those early discussions about a magical future took place around a regular old kitchen table. That's where the fifteen or so industrial designers, including Ive, Kerr, Richard Howarth, Eugene Whang, Shin Nishibori, Douglas Satzger, and Christopher Stringer, would regularly meet. "We'll sit there with our sketchbooks and trade ideas," Stringer said. "That's where the really hard, brutal honest criticism comes in."

The ID team made innumerable designs variants, which were discussed, examined, and ruled out. Inspiration struck and then vanished. At one point, Nishibori was instructed to look at what Sony was doing; as a lark, he modeled one iPhone design on the Japanese company's style, complete with a tongue-in-cheek Jony logo.

Two concepts came to the fore: One, put forward by Stringer, was inspired by the aluminum-bodied iPod Mini and came to be known as Extrudo. Made of extruded aluminum, with its hard edges and smaller screen, it had an aggressive, sharp feel. It looked a bit like a cross between an iPod and an electric shaver. Like the Mini, it could be minted in various colors of anodized aluminum. The other design was Howarth's and came to be known as the Sandwich — a rectangle with rounded edges, it was made of two sheets of plastic with a metal band running around the length of the body.

The ID group wasn't allowed to see the user interface either, so they worked on devices using stickers with cartoon versions of the apps on the screen. Perhaps unsurprisingly, given Ive's well-known affinity for aluminum, the team preferred the Extrudo model. Plus, Apple already had factories pumping out palm-size aluminum gadgets, and it would make for a less painful bridge when it came to ramping up supply. Extrudo was the first design the ID group sent to the hardware team to build out.

"We went through two different form factors," David Tupman tells me. "The first one was just like a very big iPod Mini that was literally screwed into an aluminum tube and cut out for a screen.

"We built working prototypes with electronics in it," Tupman says. "It was beautiful, as Jony makes all these things beautiful, but it just had hard edges to it." Those hard edges bothered just about anyone who tested it. Extrudo, sadly, was uncomfortable to the face, a pretty serious disqualifier for a phone.

Meanwhile, the solid-metal enclosure made it nearly impossible to deliver a signal. Two engineers, Phil Kearney and Rubén Caballero, Apple's antenna expert, had to go deliver the bad news in a boardroom meeting with Jobs and Ive. "And it was not an easy explanation," Kearney said. "Most of the designers are artists. The last science class they took was in eighth grade. But they have a lot of power at Apple. So they ask, 'Why can't we just make a little seam for the radio waves to escape through?' And you have to explain to them why you just can't."

So the team tried to accommodate both the rough edges and the radio problem. "We made books and books filled with pages of designs, trying to figure out how not to break up the design because of the antenna, how not to make the earpiece too hard and sharp, and so on," Doug Satzger said. "But it seemed like all the solutions that added comfort detracted from the overall design."

Eventually, Jobs decided to kill it. "I didn't sleep last night," he said, "because I realized that I just don't love it." He said he felt that the design didn't defer enough to the screen, that it was too masculine. "I remember feeling absolutely embarrassed that he had to make the observation," Ive said, as he agreed immediately that Jobs was right.

"Jony and Steve one day just decided, 'We need to redo this,'" Tupman says with a cheerful sigh. "So we had to kind of redo that for the second one. And it was absolutely the right thing to do. But it was a big challenge." After Extrudo was killed, the team briefly turned to Howarth's Sandwich design. But those prototypes came back too fat and ugly—ID would shelve them until the iPhone 4, which would be based on the design, once the chips and mechanics could be appropriately slimmed down.

They'd eventually settle on an earlier design, one that looks a lot like the team's first ideas. It's impossible to know where all the inspiration ultimately came from, but it's fair to say that they too ran the gamut—and we can find clues in the citations of the first iPhone's design patent. The very first citation on one of the very first iPhone design patents the team won was for a drawing board patented in 1944 by José Ugalde, a Mexican physician. There's little record left of his work besides two archived newspaper articles about a rainmaking "ionization" machine he invented shortly before his death—and his iPhone-influencing drawing board.

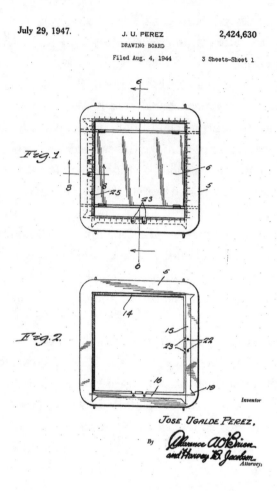

July 29, 1947.   J. U. PEREZ   2,424,630
                              DRAWING BOARD
                          Filed Aug. 4, 1944   3 Sheets—Sheet 1

Inventor
JOSE UGALDE PEREZ,
By
Attorneys

"With something like the iPhone, everything defers to the display," Ive said. "A lot of what we seem to be doing in a product like that is getting design out of the way. And I think when forms develop with that sort of reason, and they're not just arbitrary shapes, it feels almost inevitable. It feels almost undesigned. It feels like — of course it's that way, why would it be any other way?"

Engineering 101

David Tupman had the unenviable job of coordinating the mercurial pre-phone's hardware with a tiny, overworked crew. "My team,

356

which was all the electronic systems inside the phone, its RF systems, you know, the GSM systems, the Wi-Fi, the apps processor, codecs, camera, audio, and speakers and all that—that team was actually quite small," David Tupman says. "Six people? Something like that." A former Apple executive called Tupman "the hero" of the iPhone hardware effort.

Tupman hails from England, and his cheerful manner complements a boundless know-how of engineering and logistics. He'd worked on an aborted Motorola smartphone before coming to Apple to serve as a driving force in engineering the iPod. He eagerly took on the challenges of building out the iPhone. First, it was a constant battle for space—already, Jobs and the ID team wanted the phone to be as slender as possible. "We'd been doing that all the way through iPod. That was the mantra: 'Thin is in,'" Tupman says. That led to constant debates over form and function. "Can we get a battery to last long enough," Tupman says, "and get it to look like what Jony wanted it to look like?"

From the beginning, Fadell says, Ive pushed to have the headphone jack and the SIM-card slot removed from the iPhone. "We had to fight tooth and nail to make sure we didn't remove the SIM card from the very first iPhone," Fadell says.

Jony's drive to make it as thin as possible would help set the phone apart from its competitors. But it needed to work too. "I mean, we could make it an inch thick and add that to the battery life," Tupman says. "You're fighting over every micron. Every micron of thickness and square millimeter of thickness and board area, and you just try to be as innovative as you can to make it work."

Solder in the Lion

Thinness was also why the iPhone would start a trend—which many would bemoan—of shipping a phone with a difficult- or impossible-to-remove battery. "My past experience is that whenever there's a connector, that causes you a problem," Tupman says. "And so, the battery, that is your main power source. If you get any resistance or impedance in that line, and especially with 2G radio systems, you're pulling amps of current out of the battery. If you've got any impedance in that line at all, it just gives you poor performance all over," Tupman says. "So the best way of lowest impedance is a solder joint. It doesn't deteriorate over time like a connector."

He adds, "We weren't given the mission of 'make this reparable'; we were given the mission of 'make a great product that we can ship.' We don't care about removable batteries. We've never made any of our batteries removable on the iPod. So, when you're new to doing something like this, you're not tied. We were all very naive as well. It wasn't like we were phone engineers."

One of the biggest decisions they had to make was not doing 3G. The chipsets were too big and power-hungry, and they decided to prioritize longer battery life. "It was something we got knocked for, but having the Wi-Fi was a big plus," Tupman says. "No one else had really put Wi-Fi in cell phones.

"It was hard work. There were a thousand problems every day we were trying to sort out.

"Every two weeks we'd have a divisional meeting with Steve and Jony Ive and Tim Cook and all the operational teams, ourselves, myself and Tony and the other iPhone leaders. And we would just sit down and go through all of our problems." Tupman laughs. "Steve hated that. Steve hated status meetings. They would drive him crazy. He hated hearing about problems. He just wanted them all solved. You had to tell him, 'Oh, we're worried about this,' because in two weeks when you tell him, 'Oh, this thing didn't work out,' he would be like, 'Why didn't you tell me?' So you had to balance it—how much you told him so he was informed versus he doesn't get too bored and frustrated and he goes down a rabbit hole."

How Samsung Helped Build the iPhone

And one of those problems was that they didn't have a central processing unit. They didn't have a chip nailed down that could serve as the iPhone's brain. Which was sort of an important detail. "It got to a point in February 2006 where we were thinking, 'We've got to ship a product in a year, and we don't have the main processor,'" Tupman says. "'We don't even have a timeline for that main processor. How the hell are we going to do this?'"

Fortunately, the hardware team just happened to be meeting with Samsung, which made chips for the iPod, and Fadell asked them if they had anything with an ARM 11 in it. They did—it was a chip for a cable box, but the specs were right.

"So we said, 'Okay, we want to modify it, and here's how we want to do it,'" Tupman says. "'And by the way, we have to move really fast.' We said, 'We need a chip in five months.'" Chip development normally takes a year to eighteen months. "And we were trying to do this in the latest processor technology and have the first sample in five months."

Samsung was never told it was building chips for the iPhone, of course, but the iPod was already big business and Apple was an important client. "Samsung just turned the world over to make this happen," he says. "I mean, they did everything. They brought teams over to Cupertino, we were working with teams of engineers in Korea, and just getting everything done." Apple's engineers were collaborating closely with Samsung, since they weren't even finished designing the chip yet. "I mean, we're developing the spec at the same time that they're developing the chip.

"In reality, the iPhone wouldn't have shipped on that timeline if we hadn't had them helping us with that."

iWork

Tensions were running high in the Purple Dorm. Shouting matches broke out, animosities percolated, and everyone was under immense pressure. Plus there were stress multipliers of a more primordial nature.

"It stank," Williamson says. "Because we were spending all kinds of time there." There was rotting food piled up by the door; an amalgam of BO and leftovers wafted through the place. An older engineer would take breaks to go on runs and leave his sweaty clothes in the office.

Amid the stink, the demands of the project, as promised, consumed iPhoners' lives. Vacations and holidays were out of the question. So, apparently, was paternity leave.

"It was definitely intense," Williamson says. "I had recently gotten married. I had three iPhone babies. For the first one, I think I went to the hospital—then I went right back to work. I didn't take any paternity leave. And then for the other two, I took maybe a couple of days. Yeah, it was intense. Very intense."

The engineers spent night and day in the Purple Dorm, crashing there or wobbling out of Cupertino in the late night. "I remember the hallway as being dark, because so much of the time we were there was at night," one engineer said.

Those late nights and never-ending coding sessions were exhilarating to some but toxic to others. The all-consuming, embryonic iPhone eroded relationships.

"My experience of looking back and thinking about it is not a pleasant one," Grignon says. He was working every day of the week, constantly stressed, and he gained fifty pounds. "It was especially hard on the married guys," one engineer says. "There were a lot of divorces."

And it only got worse as the project drew on.

"Seating was always a problem. We had to double people up in offices, and the smell got even worse," Williamson says. "People were sleeping in there. Not official bunks. There was a couch and a cot, and it wasn't comfortable."

"There was a huge list of problems we're trying to get through," Tupman says. "And everybody's working late, all hours. All hours of the day to try to make it work. Nobody had vacation. You know, I got married in the middle of all of this. And didn't have a honeymoon until the next year." If you worked at Apple, you were on call 24/7, 365. "You don't have vacations, you don't have holidays, you don't have any of that stuff."

Perhaps the most storied event in the lore of iPhonic overworking involves John Wright, the software engineer. He had worked all morning on a Saturday, and in the afternoon, he packed up his things and started to leave. It was his son's birthday. Kim Vorath, the product manager, saw him leaving and asked if he was planning on attending a meeting that was scheduled a bit later. When he said no, she began to chew him out: "You think any of us want to be here?" she reportedly said. "I've got kids too!" They argued in the hallway, and Vorath stormed off. She slammed the door to her office so hard that the handle broke and left her stuck in the room. Scott Forstall found an aluminum bat and used it to break in and bust her out.

"Security came by," Williamson says. "Then it was right back to work."

That's just how it was at the time, iPhone team members say. Broken doors and screaming fits were barely a blip during the marathon work sessions.

"In retrospect, it's easy to measure the cost," Greg Christie says. "While you're doing it, you're just kind of doing it. It was shockingly easy to just devote ourselves completely to this thing. At the potential expense of every other part of our lives. And I can't exactly say why that is. For some period of time, this was the most important part of our lives. Not family. Not personal health. Not physical health."

★ ★ ★

Kim Vorath is another polarizing figure in the history of the iPhone; in more than one of my interviews, she was called a battle-ax—although once with the qualifier "in good and bad ways." She drove things forward, it's said, but rubbed people the wrong way. So let's pause to note the immense gender disparity on the project. For a time, there were no women at all working on the design, engineering, or development process. Dozens of men, mostly white, no women. Grignon says that eventually, the gender breakdown would come to reflect the company as a whole, which was sadly representative of the industry at the time; that is, 10 to 15 percent women. And that includes quality assurance and administrative positions. All the names on the original iPhone patents belong to men. Abigail Brody, a creative director, wasn't UI-disclosed, though she says some of her work would be integrated into the look and feel of the phone.

It could also occasionally be an uncomfortable place to work for the handful of minorities on the staff. One iPhoner overheard another discussing some rare after-work plans, saying something to the effect of "What time do you want to meet? In my hood or yours?" That drew a rebuke from one of the managers, who told him, "We don't talk like that here." When the iPhone employee, who was not white, complained, he was reportedly treated to a very weird talk from the head of the iPhone software program, who informed him that he understood where he came from because one of his most transformative experiences was seeing Public Enemy at a concert at Stanford.

Because the iPhone has proven to be such a mammoth success, it's worth considering the fact that the device was created by an almost totally male, mostly white team. It's hard to gauge the effect that any design biases exerted there—however unintentional—might have on perhaps the largest shift in personal-computing paradigms in history. And though the devices were tested by women in Apple's quality assurance department, the design and development choices were made with men's hands on the screen, and their fingerprints shaped everything from the form factor to onscreen navigation.

★ ★ ★

If things were intense before, they were about to get worse. It was October, 2006, mere months from the phone's public announcement. Many engineers had no idea that Jobs aimed to unveil the iPhone at Macworld in January 2007, but that was the plan. And he had a problem: their main chip was catastrophically buggy.

"Chips nowadays are basically software," Grignon says. "When you make a new piece of silicon, there's some dude in Korea that's gotta actually type out code, and it gets compiled into a piece of metal, silicon. Like any software program, there's bugs in there, and we hit several of those." One, in particular, "ground the entire program to a halt. There was a pretty bad bug that manifested itself between the main chip, the CPU, and the baseband, the chip that handles the phone calls."

The Samsung team had manufactured the chip per Apple's designs, and the first ones were arriving in Cupertino. It would boot up just fine, but when the engineers tried to push it, it would crash. "We weren't getting the bandwidth out of the memory that we thought we'd be getting," Grignon says.

"Steve was ready to start firing everybody," Tupman says. "That was an emergency crisis. We're two months from announcing it, and we had a major problem with our system chip." So all of the experts at Apple's disposal were called in. "Some of the best computer scientists in the world," Tupman says. "They came on board and sat down with Samsung and went into detail. 'How can we fix this, how can we get more bandwidth?'" The result of the all-hands-on-deck chip-designing spree? A working brain for the iPhone.

"Samsung did it," Tupman says. "They built a chip as fast as they've ever built a chip in a fab. Normally it's days per layer, and it's twenty or thirty layers of silicon you're trying to build. Normally it's months and months you have to wait to get your prototypes. And they were turning this around in six weeks or something crazy."

Maps

In 2006, Apple and Google were still on friendly terms. The Purple software crew wanted to use Google as its default search engine in Safari, as it was already far and away the industry standard. In a meeting with Larry Page, Jobs happened to show him the iPhone prototype.

"Larry was blown away and thought it was awesome," Williamson says. "And during that meeting he suggested that we add Maps. Steve said, 'Ah, this makes total sense.' We were a short ways away from actually shipping. So Steve said, 'We gotta add Maps.'"

Lamiraux and Williamson headed to Mountain View. "In a couple hours we hashed out a plan where we could take the core of the code they had and run it on the iPhone," Williamson says. The Apple engineers walked out with the source code for Google Maps without any formal contractual agreement—"just a handshake between Larry and Steve." Contract negotiations would take too long, they figured, and the launch was coming up; they could work out the details later. That would never happen today, of course. But then, two of the biggest tech giants could still make a casual deal to port one of the most important software programs onto one of the most important consumer devices of the era.

"We collaborated freely," Williamson says, "so they gave us the source code and we ported it and built an application around it, and we did it incredibly quickly. Like, in a couple of weeks. At this point our relationship with Google was really good." It also established the relationship between teams; the Apple crew would do essentially the same thing with YouTube. Both apps would become major selling points on a phone that, initially, was a closed system.

There ended up being three major Google products embedded on that first iPhone: Maps, YouTube, and Search. When Apple's team did return to the negotiating table, they apparently proved successful, leaving a fascinating footnote in the iPhone saga. "We generated enough money from that search field to fund pretty much the entire software development for the iPhone," Williamson says. "And then some." The entire development of the iPhone has been reported to have cost Apple $150 million—so that deal with Google was lucrative indeed.

Testing, Testing

"I have two records for iPhone, which is great, because you can never take them back," Grignon says. "The first is, I was the first person to ever receive a phone call from an iPhone. Because my team did all the software for all that."

As the chips were being fitted, the hardware clicked into place, and the software was improving at a rapid clip. The early prototypes had to be tested, of course, to make sure that they'd work in the wild.

"We had just gotten these devices out of Asia; they went to my team—the software simulators and things like that, we put the software on," Grignon says. "And I was in a meeting one day with somebody from my office, and I get this number, this phone call to my office. I didn't recognize the number, so I was like, fuck it, so I sent it to voicemail. And at the end of the meeting, I checked my voicemail, and it was my guys! They were like, 'Dude! We're calling you from a phone, this is the first call!'"

Andy Grignon had just become the first person to decline a call on his iPhone. "Instead of being this awesome Alexander Graham Bell moment, it was just like, 'Yeah, fuck it, go to voicemail.' I think it's very apropos, given where we are now."

His second record was a little less savory, though almost as apropos. "I was the first guy to actually browse porn on the phone. We had just got these devices, we had the ability to make phone calls," he says. "So we're all just, like, getting our first bit of it, of time, actually seeing it in the flesh. We're sitting in the hallway, we're all browsing the web and checking out some of the apps. I don't know what it was, but I just felt like being a fucking weirdo, and I went over to this website called Foobies, which was Fark Boobies, and I just started looking at all these booby pictures. I was like, 'Ah, check this out,' and I'm doing pinch and zoom, and we're all laughing about it. Yeah, that was the first porn site on an iPhone."

Bas Ording had gotten one of the first finalized iPhones, and, like the other testers, he was expected to use it all the time. "Which was kind of cool, but at the same time kind of sucked too, because that's your main phone, and that thing would run out of batteries or crash or the reception wasn't always that great." That, of course, was the point—figuring out where and how the phone failed in real life. "Battery life, or things you discover when you're at your desk—'Oh, we need a ringer switch,' or when the alarm goes off, there's no way to stop it," Ording says. "Little stuff like that you discover real quick."

Jobs himself was testing the phone too, which made for some interesting troubleshooting sessions. A team was sent over to Jobs's house because the CEO had found that his Wi-Fi reception was non-existent. The culprit? "It was this brick house with two-foot-thick walls," Evan Doll recalls. "One of my friends was on that team that got sent there to debug the Wi-Fi situation, and he was, like, not sure why he got sent there. He was a software engineer who didn't know that much about Wi-Fi so he just got there and opened up his laptop and started programming away, kind of pretending to do something to help the situation."

Soon all the member of the Purple team were given proto-iPhones and instructed to use them as their primary devices, with discretion, and to test them in as realistic environments as possible.

"I hate to bring this up, because I sound like a moron when I talk about it now," Nitin Ganatra says, "but I was actually testing what it was like to SMS somebody while I was driving my car! At the time, yes, I had a message to send, but the way we developed things at Apple was to live with them and mimic what other people are going to be doing with things as best you can, to try to anticipate how people are going to use these things … so part of my flimsy justification for texting people while I was driving was this, well, when you're distracted and trying to use the keyboard, how well does it work for you? You know, if you don't have time to sit there and watch every finger hit the screen, yet you're trying to use the keyboard, how is that experience compared to if you're sitting an office in a chair that's not moving.

"Now, I was doing something that's now illegal, and it's horribly irresponsible to do, and thank goodness I didn't hit any kids or hit anybody in the car while I was doing this," Ganatra says. "I guess my point is that we were learning what the impact of what this device was along with everyone else."

Macworld

The pace of work had ramped up to breakneck speeds. Macworld was coming in early January. The iPhone was, to put it mildly, not ready. It dropped calls. The software crashed. It sometimes failed to connect to networks at all.

But delaying the phone was not an option. There was too little else to fall back on. Macworld events were legendary for new product debuts, and if Apple didn't have anything substantive to show off, the company's stock could suffer. So could its reputation; despite its best efforts to prevent leaks, the rumor mills were churning that Apple was about to announce a phone, even if nobody knew what it would look or feel like.

The main chip still wasn't ready, so the software engineers had to hack around the busted version's shortcomings. They designed what's known as a "golden path" — a sequence of actions to make it appear to the hushed crowds of tech journalists that the iPhone worked seamlessly.

The Moscone Center, naturally, was on lockdown too. Security guards policed the place, and Jobs initially tried to make it so that anyone who was on-site the night before had to sleep inside the venue, an insane idea that was shot down by other executives. But Jobs was serious about keeping the demonstration top secret.

"The graphic design group, they were going to show Steve the posters, the banners, a week before," Ording says. "And he heard there were going to be posters and he killed it right away. He said, 'No, no, no, no, there's not going to be any print stuff.' Because he didn't want to risk the night before that someone at the printing press would see those posters and go, 'Oh, an iPhone.' And it was pretty impressive, because it didn't leak at all."

"At first it was just really cool to be at rehearsals at all — kind of like a cred badge," Grignon says. "But it quickly got really uncomfortable. Very rarely did I see him become completely unglued — it happened, but mostly he just looked at you and very directly said in a very loud and stern voice, 'You are fucking up my company,' or 'If we fail, it will be because of you.' He was just very intense. And you would always feel an inch tall."

Yet a single engineer was tasked with driving the twenty or so iPhone prototypes up from Cupertino to San Francisco in the trunk of his Acura.

The One Device

"I don't think it set in for me until the morning of the announcement, in January 2007, when it was already on the front page of the newspapers," Ganatra says. "When that happened, it was like, 'Oh, wow.' I had worked at Apple for quite a while up until then and had been through some big releases for Macintoshes and things like that, but none of them had appeared on the front page on the day they were announced. It was like, 'Holy shit—people had never even seen this thing and that's a story.'"

The engineers, designers, and iPhone VIPs gathered at Moscone Center on the morning of January 9, 2007. A sort of terrified excitement coursed through the place. One of the iPhoners saw Phil Schiller messing with the device in the back—apparently seeing it for the first time, he said. "It did make me wonder why Phil got to introduce the phone, since so many other people had actually worked on it."

Like, for instance, Wayne Westerman. In a major oversight on the part of Apple's PR department, the multitouch pioneer whose technology inspired and undergirded the entire iPhone project from the beginning and who had been hired by Apple in 2005 was not invited to the announcement of the product he helped build—even as Jobs took the stage and announced that Apple had invented multitouch.

Grignon had brought a flask. "It felt like we'd gone through the demo a hundred times, and each time something went wrong," he says. "It wasn't a good feeling."

Jobs paused twenty minutes into the presentation. "Every once in a while, a revolutionary product comes along that changes everything," he began, and the rest is history. He effortlessly wheeled through the demo, showing off what he believed the key functionalities would be: the phone, with special emphasis on visual voicemail, the iPod touch, with its cover flow display, and the internet communicator, with its all-grown-up web. He demonstrated multitouch, showing off the perfected inertial scrolling and pinch-to-zoom feature, which drew massive applause. He brought Google CEO Eric Schmidt onstage, opened up Google Maps, searched for a local Starbucks, clicked the store name to call—and ordered four thousand lattes. A confused barista didn't have time to respond before he hung up to rapturous laughter inside Moscone.

The iPhone had successfully captured the technology world's undivided attention. The applause would continue onto the blogs and the headlines everywhere. It was quickly dubbed the Jesus phone by Apple watchers and warily denounced by competitors. The phone's media-rich, touchscreen-based interface and its beautiful design was a hit. Grignon's team drained his flask and spent the day drinking in the city to celebrate. Abigail Brody, the creative director, says she saw some of the design ideas she'd put together for the mystery P2 project, including the big lettering and clownfish wallpaper, in Jobs's demo for the first time. She was as surprised as anyone, she says, and honored. "I did not know it would be the first iPhone."

Inside Apple, the successful launch meant that Forstall had triumphed, Grignon says. "It set the stage politically for what was eventually going to happen, which was Tony being ousted. That was foretold. You saw that in the intro, when he swiped him to delete. In the introduction, Steve is showing how easy it is to manage your contact list, right? And he's introducing swipe to delete. And he's like, If there's something here you don't want, no complicated thing, blah blah blah, you just swipe it away—and it was Tony Fadell. You just flick it, I can delete him, and he's gone. And I was like, Ahhhh, and the audience was doing this clap-clap—except for at Apple, everyone who was on the project was like, 'Holy fuck.' That was a message. He was basically saying, 'Tony's out.' Because in rehearsals, he wasn't deleting Tony. He just deleted another random contact." In our interview, I do a double take; that can't possibly be true—it seemed so cruel. "That's what Steve would do," Grignon says. "I mean, when you look at how, you know, there was a lot of foreshadowing, and he would do stuff like that. That was one of the more visible ones. That was so obvious to everybody. Everyone was like, 'Jesus, did you see what just happened?' "

In a nod to the toll the development process had taken on his employees' lives, Jobs concluded the demo by thanking their families. "They haven't seen a lot of us, 'specially in the last six months," he said. "And, uh, as I've said often, you know, without the support of our families, we couldn't do what we do. We get to do this amazing work and they understand when we're not home for dinner on time and when commitments we've made we can't keep, 'cause we gotta be in the lab, working on something 'cause the intro is coming up. And, uh, you don't know how much we need you and appreciate you. So thank you."

The thanks may have been just a bit premature, however, since the work was anything but done. There was still a six-month marathon session left before the device would ship. And the Purple team was going to need a little more help.

★ ★ ★

The day of the announcement, Evan Doll put in a request to join the iPhone team. Until then, like most of Apple, he'd had no idea that the iPhone was bubbling over.

"I remember interviewing with Scott Forstall," Doll says, "and partway through the interview, he had his iPhone sitting on the table. No one else in the world could even touch an iPhone. And it started to ring and he picked it up and showed it to me. It said: Steve Jobs calling. And he was like, 'I've got to take this, give me one second.' And he walks out, so I sit there in this conference room for like fifteen minutes, waiting for him to come back. I'm like, 'Is he fucking with me? Is this a test?' I didn't have my own iPhone, of course, to screw around and pass the time on, so I'm just like, 'Hhmmm hmmm hmmm.' " He taps on the desk. "We used to just, like, sit there in rooms and stare at the wall and think and wait for people." He laughs. "Eventually he came back."

Two weeks later, he was on the team, and he was thrown into the fire. "You look at that first phone, and the apps on the screen, and most of them had one person working on them, sometimes a fraction of a person working on them." Doll quickly became a jack-of-all-trades, helping on the clock, on mail, on whatever needed doing as the engineering team barreled down the homestretch to a June launch. Of course, there were hiccups.

"One of the engineers on the team was trying to debug an issue in the address-book app," Doll recalls. The engineer wasn't sure if the code he was writing was actually having any effect on what he was seeing on the screen. So he changed the City field in the address book to "Go fuck yourself." "He was frustrated," the iPhoner says, "and he was like, Is this even working? And then he accidentally committed that change to the repository. And that, and before he noticed it and reverted it, that was the version that went out to the carrier build that AT&T was testing out in the wild."

Before long, Scott Forstall got this phone call from the CEO of AT&T saying, "Why is my iPhone telling me to go fuck myself?" Management was not amused. "That engineer had to send an email out to the whole team apologizing for bringing dishonor to the family." A security engineer, meanwhile, had taken his pre-launch phone on a trip and had shown it to a sommelier—who turned around and published a full rundown of the new phone on an Apple-rumor website. The engineer would have been fired, but he was the only one with knowledge of some of the phone's encryption systems. "And so he also had to send out a penalty-box apologetic 'Oh, sorry, our hard work,' and blah-blah-blah."

Some of the younger engineers thought it was weird that teammates had to be punished in such a public way among the team. "Very kind of echoes-of-a-totalitarian-regime sort of public humiliation," one recalled thinking. Similar to, for instance, how factory workers had to do penance in front of their peers at Foxconn. "Yeah, there are some eerie parallels there."

The work would continue at a frantic pace for the next three months as the engineers scrambled to move the phone beyond the golden path. With the help of Samsung's tireless chip team, the custom ARM was finished and slotted in. A month after the demo, Jobs made his famous decision to switch from a plastic screen to glass and pushed Corning to churn out enough Gorilla Glass to cover the first iPhone's screens. Bugs were debugged. The address book was profanity-free. The phone stopped dropping (as many) calls.

Richardson's team killed the baby web in one swoop and successfully squeezed Safari onto the iPhone. Given the rapturous reception to the Maps demo, Jobs approved the last-minute addition of YouTube. With Google's help, the engineers had it up in running in a matter of weeks.

"We didn't realize at the time what we were doing," Lamiraux says. "I will always remember him sitting on the couch—we just had the YouTube app up and running, and he was sitting there playing with the YouTube app, and he said, 'You guys, you probably have no idea, what you're doing is more important than what we did with the original Mac.' And we were like, 'Okay. Thanks, Steve.' But I think he was right."

Launching an Icon

When the iPhone launched in June 2007, lines snaked around Apple Stores around the world. Diehards vying to be the first to own the Jesus phone waited outside for hours, even days. The media covered the buzz exhaustively. But despite all the spectacle, after a strong opening weekend—when Apple says it moved 270,000 units in thirty hours or so—sales were actually relatively slow. For now, the app selection was locked, the phone ran only on painfully slow 2G networks, and nothing was customizable, not even the wallpaper. And it was lambasted for being too expensive. Microsoft CEO Steve Ballmer famously scoffed, "Five hundred dollars? Fully subsidized? With a plan? . . . That is the most expensive phone in the world."

The hit features were probably Safari and Maps—two media-rich, multitouch-powered experiences that set the stage for everything the iPhone was capable of. That, and the touchscreen itself. That small team that had Explored New Rich Interactions starting five years before had indeed nailed their vision of how people were hoping to interact with devices. Bas, Imran, and company's groundbreaking user interface was inviting, intuitive — and addictive.

When Apple lowered its price and added the App Store the following year, the iPhone would rise to a global powerhouse. Yet it's still striking to consider how little the fundamentals have changed since that first iteration. The screen's bigger, but we all still open our phones to a grid of round-edged icons. We still rely on Safari to search, Messages to chat. We navigate by multi-fingered touch; we still watch videos on its black mirrored screen. The immediacy of the core animations still entices us to swipe, press, and tap; we still scroll through lists of information with the flick of a finger.

"It stands the test of time," Tony Fadell says. "You look at the base assumptions, and what's changed? The business model changed. Sure, better camera, better whatever. But the fundamentals have not shifted since. It's always bigger, faster. But nothing has really changed. There was no fundamental shift in the idea that would allow you to use it all. Version one was the right thing. Even though you had to iterate and kill, kill, kill — that's what happens when you get it fundamentally correct. That's what tells you you have a classic device."

Classic is one way to put. *Like water* is another.

"The impact of the iPhone has been so huge and so fast," Lamiraux says. "Compared to the Mac . . . I think the iPhone has had more impact on the life we have today. But if you think about it, this is a Mac. We took a Mac and we squished it into a little box. It's a Mac Two. It's the same DNA. The same continuity."

And that's an important point. Even the engineers who made it all possible know that they're standing on the shoulders of giants or, as Bill Buxton would say, part of the long nose of innovation. The iPhone may have seemed like a new leapfrog invention, but not only were its creators relying on a spate of technologies developed for decades outside of Apple, they were seizing on and refining a legacy long built within its ranks.

"Products like multitouch were incubated for many, many years," Doll says. "Core Animation as well had been worked on for quite a while prior to the phone. Scott Forstall, who led up the whole iPhone effort as a VP, was a rank-and-file engineer working on these same frameworks that evolved into what you use to build iPhone apps," he says. "And those were not invented in a year, or created in a year, they were created over probably twenty years, or fifteen years before the iPhone came around."

Those frameworks are made of code that's been written, improved, and recombined since the 1980s—since the days of NeXT, before the modern Apple era—by some of the same people who were instrumental in building the iPhone. "If you use any of the frameworks now, on iPhone, they have an NS prefix. Anything that has an NS prefix is NeXTSTEP code, and it pretty much is exactly the same code," Williamson says. "Now, things evolve quite a bit, and things have gotten more complex," but from NeXT to Mac to iPhone, "it's not like an unclear path; it's direct."

Apple had been banking code, ideas, and talent for twenty years at that point. "There was a compounding interest effect that was happening," Doll says. "I think that's the best way to describe it. Your bank account has been accruing interest for a while, suddenly your three percent a year, when you're twenty years in, you start to see this ramp up in the curve. I do think that was a big part of it." Another big part of it? Simple luck.

The ENRI team created a batch of interaction demos on an experimental touchscreen rig—right before Apple needed a successor to the iPod. FingerWorks came to market with consumer-friendly multitouch—just in time for the ENRI crew to use it as a foundation. Computer chips had to shrink. "So much of it is timing and getting lucky," Doll says. "Maybe the ARM chips that powered the iPhone had been in development for a very long time, and maybe fortuitously had reached a happy place in terms of their capabilities. The stars aligned." They also aligned with lithium-ion-battery technology, and with the compacting of cameras. With the accretion of

China's skilled labor force, and the surfeit of cheaper metals around the world. The list goes on. "It's not just a question of waking up one morning in 2006 and deciding that you're going to build the iPhone; it's a matter of making these nonintuitive investments and failed products and crazy experimentation—and being able to operate on this huge timescale," Doll says. "Most companies aren't able to do that. *Apple* almost wasn't able to do that."

When the right market incentives arrived at its doorstep, Apple tapped into that bank of nonintuitive investments that had been accruing interest for decades. From its code base to its design standards, Apple drew from its legacy of assets to translate the ancient dream of a universal communicator into a smartphone. It also tapped its formidable talent pool. Hundreds of people. And not just the Purple team that wrote the magic software and the iPod team that harnessed the hardware, but so many other teams inside Apple and outside it—Samsung's chip team, Corning's Gorilla Glass crew, and a long list of third-party suppliers. Carrier relations. They all worked like hell to envision, invent, and carry out the grunt work of creating the device.

Steve Jobs would routinely shout out the "Apple team" or the "great team" at demonstrations and in interviews, but he would rarely name them, unless they were executives. He told his biographer Walter Isaacson that his favorite Apple product was the Apple team—but it was the one product he was apparently unwilling to show off to the world.

"Apple was lucky to have people that loved each other so much, working on a project so key to its future," Imran says. "I can't think of another collaboration like it I've ever had."

One iPhone designer told me that he doesn't think there's a single picture of the original design team in existence. "It's like if Michael Jordan was a ghost," he told me. "There's this thing that scores and slam-dunks and wins all these games; nobody knows it actually exists." The truth is, the designers, engineers, and programmers who contribute to it know. Especially when they worked so hard they sacrificed their families, their health, everything, to create a product for Apple.

"My family did suffer, in the early days, from me not being around," Richard Williamson says. His wife died from brain cancer around the same time he left Apple. "I'm making up for that now, because I'm a full-time dad now. I cook dinner for my kids every day. It's the right thing for me, and the right thing for them. But I wouldn't exchange the experience of building the iPhone for anything."

Brett Bilbrey, who wasn't a core member of the iPhone team but who was involved in some of the research and engineering projects around it at the time, puts it like this: "I retired because of many reasons. And stress was one of them. It was a time of chaos, politics gone wild, fiefdoms. Steve was the one ring to rule them all. And people around me were dying. From heart attacks, from cancer. I do miss Apple. It was my dream job," he says, and his wife chimes in from the background, "Until it almost killed you!" His doctor, he says, gave him an ultimatum. Do these two things or risk dying—lose weight and quit. "Thirty-six people I worked with at Apple have died," he says. "It is intense."

That intensity is also likely the reason that the team that built the iPhone has since scattered to the winds. As of 2017, besides Jony Ive, none of the executive staff at Apple was seriously involved in creating the iPhone. Fadell exited the year after its launch. Scott Forstall was pushed out after the faulty release of Apple Maps in what many speculate was the culmination of long-brewing tensions with Tim Cook and Ive. Richard Williamson was fired too, despite over a decade and a half of service to the company. Burned out, Andy Grignon quit shortly after the iPhone launch. Bas Ording left in 2013 to take a job at Tesla—he was tired of spending his time defending patents in court. Henri Lamiraux retired after the rollout of iOS 7, also in 2013, for health reasons. Greg Christie, the head of the human-interaction team, left in 2014. David Tupman left that year too. Imran Chaudhri, perhaps the last father of the iPhone standing, left in early 2017.

"I don't think there's anyone left there who understands the iPhone stack from the ground up," Williamson says. In fact, the story of how the iPhone was made and who helped create it isn't even well understood inside Apple.

The iPhone project is no longer about assembling a fresh constellation of interaction ideas or inventing new ways to bring mobile computing for the masses—it's about selling more iPhones, which of course makes sense. It's business. "It's interesting to see how people perceive the company now versus then, how that has changed," one original iPhone team member says. "It's not that kind of Rebel Alliance vibe—we're Big Brother now."

Some companies might have tried to preserve the team that innovated so brilliantly together, to promote its members, or even to replicate its ingredients. But there would be only one Purple Project. At least its legacy would be formidable.

"Nobody says 'I don't know computers' anymore," Imran Chaudhri says. "That went away because of the work we did."

Iterating

The first computers were people. Skilled laborers working for astronomers and mathematicians, completing lengthy, complex calculations, often in teams, always by hand. Usually apprentices or women, they would spend days, even weeks, working out equations that could be solved today in a nanosecond with the tap of a button. But from the seventeenth century to the mid-twentieth, when these computers helped the military calculate weapons trajectories or NASA map out flight plans, the term *computers* described working people. And not only laborers, but laborers who were mostly invisible, working to benefit a man or institution that would ultimately obscure their participation.

In fact, the actual origin of computing as we know it today probably begins not with the likes of Charles Babbage or Alan Turing or Steve Jobs but with a French astronomer, Alexis Clairaut, who was trying to solve the three-body problem. So he enlisted two fellow astronomers to help him carry out the calculations, thus dividing up labor to more efficiently compute his equations. Two centuries later, six women computers programmed the ENIAC, one of the first bona fide computing machines, but they were not invited to its public unveiling at the University of Pennsylvania nor mentioned at the event. Today, the iPhone hides the fact that it contains a computer at all.

Of course, the meaning of the word has changed, but it's worth thinking about the computer — especially this, the bestselling computing device of all time — as being powered by human work. Because the iPhone, more expertly than its many predecessors, hides the immense amount of effort and ingenuity that's gone into it. As the screens get sharper, the apps get more addictive, and the phone becomes more seamlessly integrated into our daily routines, we're drifting further away from grasping computing as the work of human beings — at a time when they are in fact the work of more human beings than ever.

That's *beings*, plural. Now, Steve Jobs will forever be associated with the iPhone. He towers over it, he introduced it to the world, he evangelized it, he helmed the company that produced it. But he did not invent it. I think back to David Edgerton's comment that even now, in the age of information animated by our one devices, the smartphone's creation myth endures. For every Steve Jobs, there are countless Frank Canovas, Sophie Wilsons, Wayne Westermans, Mitsuaki Oshimas. And I think back to Bill Buxton's long nose of innovation, and to the notion that progress drives ideas continually into the air. Proving the lone-inventor myth inadequate does not diminish Jobs's role as curator, editor, bar-setter — it elevates the role of everyone else to show he was not alone in making it possible. I hope my jaunt into the heart of the iPhone has helped demonstrate that the one device is the work of countless inventors and factory workers, miners and recyclers, brilliant thinkers and child laborers, and revolutionary designers and cunning engineers. Of long-evolving technologies, of collaborative, incremental work, of fledgling start-ups and massive public-research institutions.

And all of those forces continue to shape it today. The iPhone is drawn from ideas, materials, and parts taken from every continent in the world; it was designed and prototyped in one corner, mined in another, manufactured in another still—and its influence is shipped right back to all those places and many more.

I made it a point in my interviews to ask those who worked on the original iPhone project how they felt about the device they'd unleashed upon the world—and was surprised to find a near-universal ambivalence. Most were awed by the reach of the device, by the boom of apps it begot. Most also mentioned the downsides of its constant distraction, lamenting couples eating dinner together gazing silently into their devices.

One, Greg Christie, whose dream was to make a mobile computer, has a first-generation iPhone buried under his house. "I had the hardware guys pull the battery out of the original iPhone, and it's in there along with the newspaper from that day. Picture of my family. And a note," he says. "It's encased in the base of the porch. It's kind of my life's work."

But it was the man who oversaw the software engineering for the most influential device of our time gave me the response that startled me most.

"I see people carrying their phone everywhere all the time," Lamiraux says. "I'm like, okay, it's kind of amazing. But, you know, software is not like—my wife is a painter. She does oil painting. When she does something, it's there forever. Technology—in twenty years, who's going to care about an iPhone?"

Technology is an advancing tide, he means, and even the achievements that led to something as popular and influential as the iPhone will eventually be swept away. "It doesn't last," he says. He has been writing code for decades, he says, and it's almost all been erased and replaced. "The frameworks are still there, though." That's a pretty nice metaphor for technological progress, actually. His work contributed to a larger, more permanent body, a framework that other people will build on, plug into, advance, and exploit.

"It's not like you created something, a piece of music that's going to be appreciated for a long, long time," he says. "It's just going to disappear and be replaced by something better, and be gone." If not gone, then close to invisible. A step forward, toward who knows what, in an ocean of compiled progress. That work may ultimately be unseeable, but it's also indispensable and adds to the glue that holds our improving technologies, our frameworks for interfacing with the world, together.

Computers were human once, and they always will be, to a certain extent. Because that something better is surely inching along with the help of hundreds of thousands of discoverers, engineers, laborers, designers, scientists, dealmakers, researchers, and miners. The next one device is no doubt already in the process of being pulled out of the earth.

Notes on Sources

The iPhone truly is a convergence technology, or, as computer historian Chris Garcia terms it, a confluence technology. There are so many highly evolved and mature technologies packed into our slim rectangles, blending apparently seamlessly together, that they have converged into a product that may resemble magic. Investigating the origins and inspirations of such a device was therefore a complicated undertaking, one that required making certain choices about which technologies, locations, and personalities to examine.

It meant identifying what I came to believe were the key ingredients of the iPhone — the dream of an audiovisual communicator, multitouch technology, a low-power/high-performance processor, groundbreaking user interface design, and so on — and exploring their roots. So I approached each chapter by interviewing a technology's inventors and innovators, as well as historians and analysts who study that technology, sifting through the published research and patents relevant to the field and traveling to key locales that have felt the subject's influence or contributed to its rise. On multitouch, for instance, I interviewed Bill Buxton, an early pioneer of the field, traveled to CERN to witness the environment in which a step of its evolution unfolded, and looked back through the patent filings of touchscreen pioneer E. A. Johnson. This helped give me a robust sense of the oft-overlooked history of the technology, and rendered a vital rebuke to Steve Jobs's more popular claim that Apple "invented multitouch." Speaking with Buxton, and touch innovator Bent Stumpe, helped me present a portrait of an unfathomably complex tapestry of invention, laced together by personalities with complex relationships to their work and place in history.

Meanwhile, perhaps the most crucial part of understanding the origins of the iPhone was interviewing miners, factory workers, and e-waste recyclers and repairers, which revealed difficult truths about the most ubiquitous device of our time. I was willing to "trespass" onto Foxconn's grounds because I believe it's in the public interest to better understand how the world's most ubiquitous gadget is made.

The Apple chapters are a different story. As mentioned, Apple is notoriously secretive—its strict nondisclosure policies mean that an employee who leaks can be fired on the spot. I'm told that former employees who speak to the press even after they leave stand to lose benefits (and of course, stature). So this section had to be conducted with care—I spent what felt like days on LinkedIn and writing emails, reaching out to every iPhone member I could find whose names were either listed on its primary patents or linked to the story in interviews, testimonials, and media coverage. Those still at Apple had to speak anonymously, or risk losing their jobs. Others declined to talk altogether. I thought it worth including anonymous sources here—all of whom I believe gave trustworthy testimony—given Apple's intensely secret nature. I also pored over court filings, depositions, and public testimony given especially during the Samsung copyright trials. Many, even most, of the key members of the team involved in creating the software and hardware for the iPhone have since left Apple, and I was able to interview them on the record— which is what you see here.

Teardown

To get a handle on how ubiquitous smartphones have become, I turned to research organizations and market data. Pew started tracking smartphone ownership and usage in 2011, when it estimated 35 percent of Americans owned smartphones. That number doubled in just five years. In 2007, according to ComScore figures, 9 million Americans owned smartphones. Given that there were 301 million U.S. residents then, that means smartphone ownership was just shy of 3 percent. Since then, the matter's been polled more thoroughly, of course, and Pew's 2017 report concludes that 77 percent of Americans own smartphones.

Today, Nielsen and an array of marketing firms chart smartphone screen-time habits. The 85 percent usage figure comes from the Marketing Cloud 2014 Mobile Behavior Report. The study about our perceptions of usage was led by Sally Andrews, a psychologist at Nottingham Trent University. The market-research group Informate arrived at the 4.7 hours screen-time figure. Dediu's argument that the iPhone is "the most popular product of all time" can be found at the post of the same name at Asymco's website. The 70 percent profitability figure comes from a report run in Recode, based on raw-cost findings of IHS, a UK-based market-research firm; the research was also covered in the *Independent* in "Apple's iPhone: The Most Profitable Product in History." The 41 percent figure comes from a 2014 Credit Suisse analysis. The analysis concluding that the iPhone is more profitable than cigarettes was conducted by 247 Wall Street, based on data from S&P Capital IQ, and published on *Time* magazine's website.

For historical context here and in future chapters, I interviewed Jon Agar, a historian of mobile technology, who wrote *Constant Touch,* one of the few historical surveys of the segment. I corresponded via email with David Edgerton, a historian of technology and author of *Shock of the Old,* for further context. Mariana Mazzucato's book *The Entrepreneurial State* contains an entire chapter about the iPhone and how government-backed agencies and initiatives contributed to each of its key technologies. It's a controversial book in some circles; critics argue it gives too much credit to governments and not enough to entrepreneurs. But its thesis is indisputable—the foundation for the most prominent technological products are often laid by the state as they require immense funding that only such large institutions are capable of. You'll find many examples of this throughout the book, whether it's the British navy ponying up for wireless comms systems for their fleet; CERN fostering innovations, from the web to touch-screens; or DARPA investing in artificially intelligent assistants.

Equally indispensable was Mark E. Lemley's "The Myth of the Sole Inventor," a paper published in the *Michigan Law Review* in 2012. Lemley is an esteemed patent lawyer, and the case he makes—that inventions occur both collaboratively and simultaneously, that ideas are "in the air"—is a central concept of this book. No one inventor should be credited with, nor is a single inventor capable of, crafting such an influential device. For the teardown section, I traveled to San Luis Obispo, California, where iFixit's headquarters have been refashioned from an old car dealership. In addition to my guided teardown with Andrew Goldberg, I interviewed iFixit CEO Kyle Wiens and a couple of other members of the team.

1. A Smarter Phone

To explore the roots of the smartphone, both as a concept and as a piece of convergence technology, I spoke to a number of technology histori-ans, industry vets, and science-fiction experts. These include Chris Gar-cia, curator of the Computer History Museum in Palo Alto; Matt Novak of Paleofuture; and Gerry Canavan, a professor of literature who spe-cializes in science fiction at the University of Marquette. Kristina Wool-sey, the onetime director of Apple's Multimedia Lab, and Fabrice Florin, a former executive producer at Apple, provided context.

Herbert Casson's *The History of the Telephone*—published in 1910 and available to read for free on iBooks—was a fascinating look at telephony from the same perspective we have of the iPhone today, a decade or so after its popularization. Carolyn Marvin's *When Old Technologies Were New* provided context about the dawn of the electric age. Agar's *Constant Touch* was a reference on the evolution of mobile technologies. Albert Robida's *The Twentieth Century* is a portentous look at how audiovisual technologies might evolve. Other sources include "As We May Think" by Vannevar Bush, which imagines the future of human knowledge augmentation and the memex; J.C.R. Licklider's *Man-Computer Symbiosis,* which half predicted the iPhone through a skewed lens; Norbert Wiener's *Cybernetics,* which outlines the ways that a computer control system can influence lives; and Alan Kay and Adele Goldberg's *Personal Dynamic Media,* which outlined a vision for personal computing that would set the enduring standard.

The core of the chapter is Frank Canova, who was kind enough to demo the original Simon for me and whom I interviewed at his office in Santa Clara.

2. Minephones

My visit to Cerro Rico was conducted through a local mine-tour company, which hosted my fixer and me on a personal expedition. We spoke to the miners we encountered on the site and a local *colectivo* boss who arranges distribution to smelters, and I interviewed former miners like Ifran Manene. He, among others, confirmed that ore from Cerro Rico was sent to the smelters listed by Apple. The top industry trade group, ITRI, confirmed the flow of tin from Potosí to Apple smelters.

Investigations into the mining practices fueling the tech trade have been carried out by the *Washington Post,* the *Guardian,* the BBC, and many other organizations, and their invaluable work informed this chapter. BBC Future reporter Tim Maughan provided crucial insight about the rare earth extraction site in Baotou. My former colleague Wes Enzinna's "Unaccompanied Miners" piece for *Vice* magazine, about the child-mining epidemic in Bolivia, was a valuable source of reportage as well. NPR's segment "Thousands of Children as Young as 6 Work in Bolivia's Mines" confirms the age statistic.

David Michaud, the metallurgist who runs 911 Metallurgist, a mining consultancy, arranged to have my iPhone 6 pulverized and analyzed by chemical scientists and prepared a report about the results. We will publish those results concurrently with the book. Michaud was also interviewed about the results. The calculations are his, and drawn from available data about mining operations that can be considered industry standard—not the mines Apple gets its minerals from in particular, most of which are undisclosed. He also notes that the "20.5 grams of cyanide" figure "depends on the source of gold. This is an average number. It may vary from 5 to 60 grams."

Figures and information about mining and pollution were drawn from the EPA, the World Bank, and UNICEF. The journalist Elizabeth Woyke's *The Smartphone: An Anatomy of an Industry* provided a good overview of the raw materials that make the technology possible. Jack Weatherford's book *Indian Givers: How the Indians of the Americas Transformed the World*, provided context about the history of Potosí and its indigenous population.

3. Scratchproof

The story of how Corning came to make the iPhone's glass was mentioned in Isaacson's *Steve Jobs*, detailed in Fred Vogelstein's *Dogfight*, and further fleshed out in Bryan Gardiner's excellent *Wired* cover story "Glass Works." Gardiner was kind enough to sit for a phone interview and supply further details. The quotes from Corning's Kentucky plant workers come from a 2012 NPR story called "Small Kentucky Town Makes High-Tech Glass Amid Bucolic Farmland." Further information came from Daniel Gross and Davis Dyer's *The Generations of Corning,* a history of the company that includes details about the Project Muscle and Pyroceram programs. Finally, Corning's vision for a world filled with glass products was detailed in its bit of design fiction, *A Day Made of Glass,* a video released in 2012.

4. Multitouched

An entire book could and should be written about the history of touch technology. I interviewed Bill Buxton, the pioneer who used the term in a paper he published at the University of Toronto in 1984. His compendium of touch technologies, "Multi-Touch Systems I

Have Known and Loved," is one of the best resources on the topic anywhere. I also reviewed the patents for E. A. Johnson's first touch-screen, the early CERN yellow papers, as well as Apple and Finger-Works' patents. I traveled to CERN and interviewed Bent Stumpe and David Mazur, and a number of other current employees of the consortium. Stumpe gave me one of the earliest touch screens he prototyped — capable, he says, of multitouch.

I explored the work of the early synthesizer pioneers and listened to a lot of theremin and piano duets performed by Clara Rockmore and Sergey Rachmaninoff. James C. Worthy's biography of William Norris, *Portrait of a Maverick,* provided details about CDC and Nor-ris's decades-long advocacy of touch technologies.

To get to the heart of Wayne Westerman's story — Apple wouldn't make him available for an interview — I interviewed Ellen Hoerle, his older sister and only living nuclear-family member. Wayne's viv-idly written 1999 dissertation on multi-finger gestures was essential to this chapter, and surprisingly fun to read. I plumbed early inter-views with his alma mater's newspaper, the *New York Times,* and the *News Journal,* which are where any quotes attributed to him origi-nate. Jeff White, the erstwhile FingerWorks CEO, gave an interview to Technical.ly/Philly, which quotes are drawn from.

As a nontouch aside, it's also worth noting that Tim Berners-Lee built the World Wide Web using a NeXT Cube — the computer made by the company Steve Jobs founded after getting fired from Apple.

5. Lion Batteries

SQM organized the tour of their facility in Atacama and allowed us to stay on-site so that we could visit both Salar de Atacama, where the lithium is harvested, and Salar de Carmen, where it is refined and prepared for distribution (I paid for the travel and the rest of the lodgings). Enrique Pena, the evaporation ponds superintendent, was interviewed about the process, and additional details came from

Claudio Uribe. David Michaud provided context. Data about the amount of lithium sold was supplied by SQM. I also traveled to Salar de Uyuni, the largest lithium deposit in the world, in nearby Bolivia, but development has not yet begun seriously there and those adventures will have to wait for another story.

For historical context about the history of the lithium-ion battery, I interviewed Stan Whittingham and John Goodenough, the two godfathers of lithium-ion-battery technology. Steve LeVine's *The Powerhouse* provides a nice history of the modern battery, and information drawn from the earlier chapters especially informs my own on the topic. *New Scientist*'s story about the Unknown Fields expedition to Bolivia, "Lithium Dreams," was another valuable resource.

6. Image Stabilization

When I was looking for an iPhone photographer to follow, I hoped to find someone who'd gotten an early jolt in his or her career thanks to opportunities provided by the device and whose photos uniquely reflected the new style of shooting. David Luraschi was perfect. Interviews with Brett Bilbrey (over phone and email) and Dr. Oshima (via email) helped me understand the advances in camera technology that led to the modern smartphone camera. Some additional quotes are drawn from a previous interview with Oshima archived at the Japan Patent Office. A *60 Minutes* story about Apple's camera factory provided the figures and data about the company's current camera operations.

7. Sensing Motion

A visit to the Musée des Arts et Métiers kicked off this chapter; in addition to housing the famed pendulum, it's also home to the Jacquard loom, Pascal's calculator, and other early computer ancestors. Sid Harza helped explain MEMS tech to the humble layman— myself—while Brian Huppi explained how Apple developed the sensors here, and Brett Bilbrey shed more light on sensor development in general. The articles by *Economist* writer Glenn Fleishman on GPS were a useful reference. For accelerometers, Patrick L. Walter's "The History of the Accelerometer," published in *Sound and Vibration*, provided exactly that. Arman Amin's thread about the motion coprocessor can be found on Reddit at r/Apple.

8. Strong-ARMed

There's no way to cram the entire history of computers or computer processors into a single chapter, so consider this the highlight reel of two of the most crucial events: the birth of the transistor and the establishment of Moore's law. It's interesting to note that the transistor kicked open the door for the cell phone, which was created almost immediately after the former's discovery at Bell Labs. James Gleick's *The Information* offered a historical framework, and Walter Isaacson's *The Innovators* provided context for the rise of the transistor.

The Apollo Guidance Computer had 4,100 NOR gates, each with three transistors—that's 12,300 transistors. According to Paul Ledak, a former IBM microprocessor exec, that means that as of 2015, the iPhone 6 had 130,000 times more transistors than the Apollo system. The single-transistor hearing aid is described at the online Semiconductor Museum.

I interviewed Alan Kay at his Brentwood home; he suggested I read Neil Postman's *Amusing Ourselves to Death,* which I did, and I recommend you do too. It's a trenchant critique of our entertainment-addled culture, now more relevant than ever—but I digress. I interviewed Sophie Wilson over FaceTime and she graciously allowed me to take up far more of her time poring over the early beginnings of ARM than I should have.

Ryan Smith of Anandtech gave me crucial context for understanding Apple's chip-development game, and an interview with industry analyst Horace Dediu yielded insight into the rise of the app economy from an insider's perspective. David Edgerton and Adam Rothstein helped provide historical context in email correspondence, pointing out that the app economy could be a new name for an old series of services. I interviewed Joel Comm by phone to get an example of an early app success.

Former Apple employees on Team iPhone—Henri Lamiraux, Nitin Ganatra, and Andy Grignon chief among them—provided insight into how and why the App Store was resisted and then rolled out. Interviews with Muthuri Kinyamu, Erik Hersman, Nelson Kwame, and Eleanor Marchant, among others, helped me understand how the mobile app shaped Kenya's tech industry and start-up scene.

9. Noise to Signal

An interview with Jon Agar provides this crash course on the history of wireless networks. The core papers of the ALOHAnet project are available for free online and offer a peek into the birth of Wi-Fi. Christoph Herzig of Philips offered insight into the booming of distributed networks and touted SmartPoles as the future, though I didn't have space to include them. In 2012, ProPublica and PBS did an excellent series on the human toll that tower maintenance takes. Tower deaths are tracked and updated by Wireless Estimator. The smartphone ownership data comes from a 2016 comScore report.

10. Hey, Siri

The backbone of the Siri chapter is a lengthy interview conducted with Tom Gruber, Apple's head of advanced development for Siri. Artificial intelligence is obviously a loaded topic—I attempted to approach it through the lens of what Siri actually does, or tries to do. The first stop on any AI reading list is Alan Turing's classic "Computing Machinery and Intelligence." Additional research concerned the Hearsay II papers. The Oral History Collection at the Charles Babbage Institute is a great resource, and the interview conducted with Raj Reddy is no different; it provides a fascinating look at the life of one of the first AI pioneers. I also drew from published talks Reddy has given. The figures about Siri and the number of requests it receives are published by Apple and have not been independently corroborated.

11. A Secure Enclave

Def Con is fertile ground for anyone looking for a crash course on cybersecurity; I spent the days in Las Vegas hanging out with hackers and interviewing early iPhone jailbreakers and analysts. I also attended Black Hat, where I heard Apple head of security Ivan Krstić's talk about the secure enclave, something that baffled even the pro cybersecurity reporters around me. Further information, expertise, and context came from interviews with security expert Dan Guido, iPhone Dev Team jailbreaker David Wang (@planetbeing), and Ronnie Tokazowski, who works for PhishMe. Apple has published fairly detailed descriptions of how the secure enclave works, but few can confirm *exactly* how it does what it says it does. Some background comes from David Kushner's profile of George Hotz, aka Geohot, for the *New Yorker,* and a catalog of the effects of Cydia's influence on iOS was documented in Alex Heath's article "Apple Owes the Jailbreak Community an Apology."

12. Designed in California, Made in China

There's been a lot of great reporting about labor conditions in China's electronics factories, not least of which the *New York Times'* Pulitzer Prize–winning series on the iEconomy by Charles Duhigg, Keith Bradsher, and David Barboza. China Labor Watch is an invaluable resource, as is the labor-academic group SACOM. I interviewed Li Wang with a translator in the summer of 2016. Liam Young of Unknown Fields Division provided context concerning the supply chain and working conditions in China in a meeting before I departed.

My fixer and translator, Wang Yang—we've chosen to use a pseudonym to protect her identity—was a tremendous help in getting factory workers to talk. She's a big reason we spoke to a couple dozen sources over the course of a handful of visits. We visited Foxconn's Longhua and Guanlan and Pegatron's Shanghai factory, as well as supplier factories such as TSMC, the chip fabricator. Of the Foxconn employees, Xu, Zhao, and their friend were the most candid, but many factory workers were willing to speak to us outside the gates, at lunchtime noodle shops, and at the local market. From these interviews, combined with research from the above sources, I feel confident I was able to capture a solid snapshot of the state of play at China's electronics factories.

I did in fact sneak into Longhua by virtue of my having to use the bathroom—this is potentially a crime of trespassing in China, but I feel it is justified due to the history of abuse and tight media controls by Foxconn. It seemed to me it would be a public service to get a fair and un-spun image of the factory.

The number of steps necessary to produce a phone was documented by ABC's *Nightline*—though that was in 2012, and the number is likely considerably greater today, as the devices have only continued to become more complex. Suicide victim Xu Lizhi had his poems, including "A Screw Fell to the Ground," collected and published in the *Shenzhen News*.

The innovations in Ford's assembly line were detailed in David Hounshell's *From the American System to Mass Production, 1800–1932*, and that book is the source of the "Pa" Klann quotes. Stephen L. Sass's history of materials science, *The Substance of Civilization*, posits that mass production began millennia ago.

13. Sellphone

To understand why Apple is so popular, you have to understand why it's so good at creating spectacle—and you have to go to an Apple Event. Like rock concerts for products, they somehow get your blood running, even if you're fully aware you're watching a well-produced infomercial. David E. Nye's *The American Technological Sublime* helped me understand this phenomenon; beyond the Wright brothers, he looks at Edison and the Hoover Dam.

The *Atlantic*'s Adrienne LaFrance, who edited the technology section for years, wrote about the difficulties of covering secretive technology companies for Nieman Journalism Lab. Our phone interview interrogated that phenomenon. The top editorial staff of *iPhoneLife* magazine participated in an interview call during which I tried to suss out what it was like to write about the iPhone daily for a living. In a phone interview, Cory Moll detailed what forming a union at Apple was like. Mark Spoonauer provided context as a tech editor and longtime vet of Apple Events. I interviewed a couple dozen Apple Store employees, but covertly, which is why I didn't publish any of their quotes in the text—retail reps aren't allowed to speak to the press. I also interviewed perhaps a dozen customers waiting in line on launch day.

As for the Tim Cook email episode—an Apple PR rep confirmed that Tim had opened my email and forwarded it on; she said he had read it first. Streak's representatives told me that their technology could "very accurately" determine which device a person used to open an email. There are still other explanations, of course—the PR rep didn't have the right information, Cook was using a VPN that outsourced his traffic, or his emails get outsourced somewhere that uses Windows.

14. Black Market

Little could be confirmed about the iPhone black market I visited in the summer of 2016—no one would speak to journalists, though its sheer size belied the notion that it was some secret operation. However, Adam Minter, an expert on e-waste and secondhand markets who was interviewed for this chapter, returned to the same location months later. The entire thing was gone. Much good reporting has been done on the woes of Guiyu, led by Basel Action Network's research. A *60*

Minutes segment tracked U.S. waste there in 2008. Estimates of waste flows were sourced from the United Nations University.

i–iV

The first two Apple sections, i and ii, are based primarily on interviews with the team responsible for carving out the interaction paradigms that formed the foundation of the iPhone—the user interface, the multitouch software, the early hardware. I conducted interviews with Bas Ording, Imran Chaudhri, Brian Huppi, Joshua Strickon, and Greg Christie, in addition to other members of the original iPhone team on background. Further details and quotes from Jony Ive were taken from Walter Isaacson's *Steve Jobs,* Leander Kahney's *Jony Ive,* and Brett Schlender's *Becoming Steve Jobs.* Steve Jobs "misremembered" the iPhone's touchscreen genesis in a Q-and-A hosted by Walt Mossberg and Kara Swisher at their annual D: All Things Digital conference.

As with the previous roman numbered sections, most of chapters iii and iV were sourced from interviews with original iPhone team members and anonymous Apple employees, previous research and reportage, and court- and FOIA-obtained documents. Among Apple personnel interviewed on the record were Bas Ording, Imran Chaudhri, Richard Williamson, Tony Fadell, Henri Lamiraux, Greg Christie, Nitin Ganatra, Andy Grignon, David Tupman, Evan Doll, Abigail Brody, Brian Huppi, Joshua Strickon, and Tom Gruber.

Quotes were drawn from the Apple/Samsung trial of 2012, when Phil Schiller and Scott Forstall took the stand. Books that provided extraordinarily useful detail, research, and background were *Dogfight,* by Fred Vogelstein; *Steve Jobs,* by Walter Isaacson; *Becoming Steve Jobs,* by Brent Schlender; *Inside Apple,* by Adam Lashinsky; and *Jony Ive,* by Leander Kahney. Quotes attributed to Jony Ive, Steve Jobs, Mike Bell, and Douglas Satzger were drawn from those sources. John Markoff's *New York Times* reporting and Steven Levy's book *The Perfect Thing* and his work in *Newsweek* were used for reference.

Sales figures cited are provided by Apple unless otherwise stated.

Acknowledgments

A key theme of this book is that little progress is possible without deep collaboration and sustained collective effort—nothing could be truer about writing this thing too. It simply wouldn't have happened without the support of family, friends, colleagues, and even, sometimes, near-strangers. The book about what made the one device possible would not be possible without any and all of them. In a way, we *all* wrote this book.

First and foremost, my incomprehensibly supportive wife, Corrina: Not only did she make huge sacrifices to ensure I could finish this on a truly insane timeline, but she was a powerful critic, editor, and idea-generator for the book itself. Her thoughts on where I should take the thing next were often better than mine. She may also be the only person who's more sick of hearing about the iPhone than I am. I can't thank her enough. I also want to thank my one-year-old son, Aldus, mostly for existing. Knowing he'd read this thing someday, or at least mainline the data into his cranium through the next one device, made me want to make it better.

Thanks to Eric Lupfer, who is certainly the best agent I've ever met—but also a sharp editor and thinker; without him, this thing wouldn't exist. And to my editor, Michael Szczerban, whose thoughtful edits helped cohere this slab of a thing into a proper book, and who only belted out a single expletive when I told him the first draft was going to be 200,000 words. The whole team at Little, Brown—Ben Allen, Nicky Guerreiro, Elizabeth Garriga, my science-eyed copyeditor, Tracy Roe, and everyone else—I should add, has been wonderful.

ACKNOWLEDGMENTS

A huge thanks to my friends and colleagues at *Motherboard*—large chunks of this book would not have happened without their know-how, assistance, and connections. Jason Koebler went so far above and beyond I actually had to tell him to stop helping me at one point—he used vacation days to accompany me to Chile and Bolivia, acted as a translator and fixer, and arranged our visit to Cerro Rico, and beyond all that, his insights, ideas, and reporting on technology were and are an invaluable resource. Lorenzo Franceschi-Biccherrai, one of the best cybersecurity reporters anywhere, stepped away from the mainframe long enough to lend me his know-how, expertise, and spare hotel bed at Def Con. He hacked into my computer to give the security chapter a close read, and if I got anything wrong, it's because he stole my password and changed it. Thanks to Nicholas Deleon, who shared his Apple contacts and insights into the consumer tech sphere with me, and is the reason I was able to get into the Apple Events and discover the joys of corresponding with Apple PR. I hope they still talk to you after this!

Thanks to Wang Yang, my fixer and translator in China, whose gumption and enthusiasm got us further than I ever would have alone, and to Eleanor Marchant, a gracious and knowledgeable guide to Nairobi's tech scene.

Thanks also to Claire Evans and Keith Wagstaff, whose close readings of early chapters and the final manuscript provided insight and inspiration. And to Alex Pasternack and Michael Byrne too, whose knowledge of science, technology, and geopolitics made them indispensable informal fact-checkers. Same with Jona Bechtolt, whose encyclopedic knowledge of Apple, thoughts on the industry, and eagerness to land an iPhone 7 on launch day helped me understand Appledom that much better.

Thanks to my parents, Tom and Sharon, who pitched in to watch the baby, offer words of encouragement, and continued to be the best support system a grown kid could ever hope to have, and to my brother, Ed. Thanks to Tim and Teresa Laughlin, my parents-in-law, who went above and beyond, time and again, offering support whenever I was in the whirlwind of a deadline. And to my wonderful grandparents Joan and Al, who were kind enough to let me use their Palo Alto home as a base whenever I was in Silicon Valley. Julie Carter and Chip Moreland, thanks for the couch, the dive bar, and the dumplings. To Mike Pearl, for *almost* letting me take apart your iPhone. Thanks to Brian Parisi, for talking shop, and Koren Shadmi, for help with early art. Thanks to Nick Rutherford and Jade Catta-Pretta, for your ears, support, and putting me up in New York. Thanks to Alexis Madrigal and Geoff Manaugh for reading the manuscript, to Tim Maughan, Robin Sloan, and Liam Young for sharing notes and talking iPhone.

Thanks to the whole crew at iFixit, especially Kyle Wiens and Andrew Goldberg, who helped kick off this journey and provided essential resources throughout. To Adam Minter, my fortuitous guide to iPhone junk. To Yoni Heisler, who was kind enough to share court documents he'd pried out of PACER. To Fred Vogelstein, who shared notes from his own amazing reporting on Apple. To Bryan Gardiner, for talking Gorilla Glass with me. And to Ashlee Vance, for offering some words of criticism and encouragement when he certainly didn't have to — and ultimately kicked my ass into writing a better book.

Thanks also to Eric Nelson, who helped get the ball rolling.

Finally, I want to thank the many interview subjects, especially those who risked their jobs or station to help me tell a truer story about the one device, and those who do strenuous, dangerous, and unappreciated work to bring it to life.

Index

About the Author

Brian Merchant is an editor at *Motherboard,* VICE's science and technology outlet, and the founder of *Terraform,* its online fiction outlet. His work has appeared in the *Guardian, Slate, Fast Company, Discovery, GOOD, Paste, Grist,* and beyond. To trace the story of the iPhone, he traveled to every inhabited continent, from the Bolivian highlands to the megalopolis of Shenzhen, using the "one device" to document the effort. He took 8,000 photos, recorded 200 hours of interviews, tapped out hundreds of notes, and had dozens of FaceTime sessions with his family back home. He went through three different iPhones: an iPhone 6, whose screen was broken and repaired three times, a black-market 4S that he bought in China but was stolen in Chile, and an iPhone 7 he snapped up on its launch day.

THE ART OF

FLOURISHING